HISTORY OF IDEAS
IN
ANCIENT GREECE

This is a volume in the Arno Press collection

HISTORY OF IDEAS
IN
ANCIENT GREECE

Advisory Editor
Gregory Vlastos

*See last pages of this volume
for a complete list of titles*

THE FRAME OF THE ANCIENT GREEK MAPS

BY
WILLIAM ARTHUR HEIDEL

ARNO PRESS

A New York Times Company

New York / 1976

Editorial Supervision: EVE NELSON

Reprint Edition 1976 by Arno Press Inc.

Reprinted from a copy in
 The Princeton University Library

HISTORY OF IDEAS IN ANCIENT GREECE
ISBN for complete set: 0-405-07285-6
See last pages of this volume for titles.

Manufactured in the United States of America

Library of Congress Cataloging in Publication Data

Heidel, William Arthur, 1868-1941.
　The frame of the ancient Greek maps.

　(History of ideas in ancient Greece)
　Reprint of the 1937 ed. published by American
Geographical Society, New York, which was issued as
no. 20 of the society's Research series.
　　1. Classical geography.　2. Earth--Figure.
I. Title.　II. Series.　III. Series: American Geographical Society of New York. Research series ; no. 20.
GA213.H4　1975　　　526'.0938　　　75-13271
ISBN 0-405-07312-7

THE FRAME OF THE
ANCIENT GREEK MAPS

AMERICAN GEOGRAPHICAL SOCIETY
RESEARCH SERIES NO. 20
J. K. WRIGHT, *Editor*

THE FRAME OF THE ANCIENT GREEK MAPS

With a Discussion of the Discovery of the Sphericity of the Earth

BY

WILLIAM ARTHUR HEIDEL

Wesleyan University
Middletown, Connecticut

AMERICAN GEOGRAPHICAL SOCIETY
BROADWAY AT 156TH STREET
NEW YORK

1937

COPYRIGHT, 1937
BY
THE AMERICAN GEOGRAPHICAL SOCIETY
OF NEW YORK

THE LORD BALTIMORE PRESS
BALTIMORE, MD.

To
M. M. H.
W. C. H.
M. P. H.
*to whom I owe more
than I can repay*

CONTENTS
CHAPTER | PAGE
PREFACE ix
INTRODUCTION 1

PART I
The Frame in Relation to the Flat Disk Earth

I THE ORIGINS OF THE FRAME......................... 7
  Early Astronomico-Geographical Observations...... 8
  The Frame of the Earliest Maps................... 11
  Dependence of Herodotus on the Ionian Maps...... 20
  Herodotus on the Nile and the Ister.............. 22

II THE SOUTHERN LIMIT OF THE FRAME................ 26
  India and Ethiopia............................... 26
  Africa 28

III THE NORTHERN LIMIT OF THE FRAME................ 31
  East of the Ister................................ 31
  The Course of the Ister.......................... 34
  The Headwaters of the Ister...................... 36

IV THE WESTERN AND EASTERN LIMITS OF THE FRAME AND
  THE IONIAN EQUATOR............................. 45
  The Western Limit................................ 45
  The Eastern Limit................................ 47
  Origins of Herodotus' Geographical Sketch of Asia.. 50
  The Ionian Equator............................... 53

V OBSERVATIONS ON WHICH THE FRAME WAS BASED..... 56

PART II
The Spherical Earth in Relation to the Frame

VI THE SUPPOSED EARLY DISCOVERY OF THE SPHERICITY
  OF THE EARTH................................... 63
  Character of the Source Materials................ 63
  Misleading Ideas Probably Due to Posidonius...... 66
  The Sphericity of the Earth Unknown to the Early
    Philosophers 67
  The Claims of Parmenides......................... 70
  The Claims of Pythagoras......................... 72
  The Claims of Archelaus.......................... 77
  Sphericity of Earth Probably Unknown Before End
    of Fifth Century Before Christ................. 79

CONTENTS

CHAPTER		PAGE
VII	PLATO AND ARISTOTLE AND THEIR SOURCES	81
	Plato on the Sphericity of the Earth	81
	Aristotle on the Sphericity of the Earth	85
	Possible Origins of the Concepts of Plato and Aristotle	92
	Philolaus	93
	Archytas	94
	Eudoxus of Cnidus	95
VIII	DETERMINATIONS OF GEOGRAPHICAL POSITIONS IN THE PERIOD AFTER ARISTOTLE	103
	The Geographical Position of India	104
	Philo's Observations on the Red Sea and the Nile	105
	Observations of Pytheas in the North	106
IX	DICAEARCHUS	110
	The Median Axis of the "Oikumene"	111
	Did Dicaearchus Estimate the Circumference of the Earth?	113
X	ERATOSTHENES	122
	Eratosthenes' Map	122

CONCLUSION

XI	THE ENDURING FRAME OF THE IONIAN MAPS	131
INDEX		137

ILLUSTRATIONS

FIG.
1. Sketch map illustrating the probable Greek concept of the *oikumene* before the time of Eratosthenes ... 6
2. The parallelogram of Ephorus ... 17

PREFACE

This book is not an attempt to prove a thesis. The problem it discusses is one that must present itself to every serious student of Greek thought, especially if he feels drawn to the Ionians, its earliest representatives in the field of science; for geography was their peculiar province, all the leading geographers down to the time of Aristotle being Ionians. Moreover, geography, as they conceived it, was from the first bound up with history and not with political history only, which is acknowledged to be the creation of the Ionians. Their entire outlook was historical in the widest sense, differing therein essentially from that of the western thinkers, who approached their problems by direct observation or by way of logical analysis. Having this historical bent, the Ionians showed a marked respect for tradition, rarely breaking with it completely but interpreting it as reason seemed to require. There is, therefore, in the Ionian school an agreement in fundamentals and an obvious continuity that shows a wholesome conservatism while allowing ample scope for innovation and development to the individual thinker. It makes for the establishment of a rational tradition such as one discovers in the history of Greek cartography.

For many years my chief interest has lain in the study of a considerable group of special traditions in the field of Greek thought. Everywhere I have found the same continuity and adherence to the essential features of the prototype, even where, as in Plato and Aristotle, there is evident a conscious, though unsuccessful, effort to combine the analytical approach with the historical.

The present study, like others I have undertaken, has occupied my thought for twenty years or more. The fundamental significance of the tropics of Herodotus, of the parallelogram of Ephorus, and of the location of the Ionian equator by Hippocrates and Polybius, I recognized quite as long ago. The data in detail necessary to fill out the picture were accumulated gradually as other interests directed my reading. The problems connected with the theory of the sphericity of the earth, especially as they relate to the Pythagoreans and to the influence of Posidonius on the final form of the doxographic tradition, have also long been one of my chief concerns, and my conclusions represent my mature judgments. As regards the character and extent of the influence of Posidonius I differ from most scholars, but I believe that in this particular I am only advancing toward completion the work of my revered teacher, Hermann Diels.

The study, in essentially its present form, was completed three years ago, awaiting publication, which has now been made possible by the American Geographical Society of New York, to the authorities of which my warmest thanks are duly given.

W. A. H.

Middletown, Conn.
Easter, 1937

INTRODUCTION

Anything of common occurrence soon becomes commonplace and passes without awaking interest. At some time, however, all these commonplace things were new, and those that were created by man originated in an interest and served a need. If one examines them closely, one may learn much regarding the life and outlook of the men who first called them into being; and generally one will find that, though more or less changed in non-essentials, the original type endures with astonishing persistence, because the outlook, the interests, and the needs of man remain essentially the same. Few things are more commonly met in the homes of the intelligent than maps; and they are doubtless frequently consulted. But how many give even a passing thought to the question how are they made? How the first maps came to be drawn very few would think to inquire.

The maps of the ancient Greeks have a special claim to our interest. Unfortunately not one remains to us; but there are extant in ancient literature sufficient indications to enable us to reconstruct them in outline and to discover the interests and principles that informed them. As will be seen, the type was fashioned in the sixth century before Christ, and that type, though altered in detail, persisted throughout ancient and even into modern times, despite the revolutionary scientific discoveries of later centuries. The maps with which we are concerned purported to be general, presenting an outline of the earth as it was then conceived. They reveal, therefore, the state of knowledge attained by the Greeks respecting the larger features of the earth and the conceptions they were able to frame regarding regions that lay beyond the limits of their experience.[1]

[1] See Aristotle, *Meteorology*, 350 a 15ff.; Strabo, II, 5, 11.

2   THE FRAME OF THE ANCIENT GREEK MAPS

An undertaking as ambitious as a world map was of course not without antecedents. Long before the first general map was attempted, we may be sure, there were conceptions of the earth, or of its parts, set forth in one way or another, which implicitly contained the schemes that were later to characterize the work of serious geographers. Whoever, for example, reads Homer sympathetically will discover approximately where the poet's knowledge ends and where he gives the reins to his imagination. The limits of *terra cognita* will gradually expand, but much of the picture will long persist; for that is the way the human mind works in every field. Herodotus [2] ridiculed the early maps because they represented the circular earth as surrounded by the River Oceanus, which he knew was part of the Homeric tradition; one might with equal justice ridicule Herodotus' own conception of the inhabited earth encircled by deserts. But one need not think only of written accounts, for there undoubtedly existed graphic sketches that were known to the first designers of general maps. The shields of Achilles and Herakles, as described by Homer [3] and Hesiod,[4] must have been suggested by things the poets had seen. Though they do not take the form of a map, they have something in common with later representations of the earth. Moreover, Homer [5] more than once gives hints of recognized sea routes and of instructions for sailing that must have included specifications of direction as well as of distance, such as a seafaring people must of course have possessed; and it may be safely taken for granted that such instructions, even when orally given, would be accompanied by

---

[2] II, 23; IV, 36.
[3] *Iliad*, XVIII, 478ff.
[4] *Scutum*, 139-317.
[5] *Odyssey*, III, 168ff.; IV, 389; X, 539. See Felix Jacoby, "Hekataios," in *Paulys Real-Encyclopädie der klassischen Altertumswissenschaft*, new edit., Vol. 7 (edited by Georg Wissowa and Wilhelm Kroll), Stuttgart, 1912, cols. 2686-2689.

INTRODUCTION 3

graphical sketches. Nor were the Greeks, presumably, the first to use such sketches: from Babylonia there has come down to us a sketch map of the Mesopotamian lands [6]; and in Egypt there have been found surveyors' or engineers' plots of restricted areas, which might be indefinitely extended. Whether there were graphical representations of entire nomes or of the Nile Valley of early date we do not know [7]; if such existed, we may imagine them to have been symbolical or diagrammatic rather than in any sense realistic.

What seems to have distinguished even the earliest Greek maps, aside from their purpose to include the whole earth, is the frame enclosing the habitable earth. How this frame originated and was developed will be discussed in the pages that follow. In the first part of the study evidence will be set forth regarding the frame adopted in the earlier maps dating from the time when the earth was regarded as a flat disk. In the second part we shall examine the origin of the Greek belief in the sphericity of the earth and shall see how the frame of the maps was brought into conformity with this doctrine.

[6] See A. T. Olmstead, *History of Assyria*, New York, 1923, p. 596; Bruno Meissner, "Babylonische und griechische Landkarten," in *Klio*, Vol. 19, 1925, pp. 97-100. The map, like those of early Greece, shows an encircling ocean ("bitter water").

[7] See F. W. von Bissing, "Ägyptische Weisheit und griechische Wissenschaft," in *Neue Jahrbücher für das klassische Altertum, Geschichte und deutsche Literatur*, Vol. 20, 1912, p. 86. In my opinion this learned scholar is even too much inclined to give credence to Eustathius (Carl Müller, *Geographi graeci minores*, Paris, 1882, Vol. 2, p. 214) and Apollonius Rhodius (*Argonautica*, IV, 272ff.), who say that Sesostris set up maps of the world among the Scythians. The reference to the Scythians shows that we are dealing with fiction. Eduard Meyer (*Set-Typhon: Eine religionsgeschichtliche Studie*, Leipzig, 1875, p. 30) quotes an inscription of Thothmes I, of the Eighteenth Dynasty, "König der beiden Länder, der beherrschen will die Sonnenwende (*sent n aten*), die nördliche und die südliche, mit seiner Hand." What this may mean I have no idea; for the Egyptians, though they might well have known the northern, certainly had neither knowledge nor a conception of the southern tropic. Clemens Alexandrinus (*Stromata*, John Potter's edit., Oxford, 1715, Vol. 2, p. 671) shows a late (Greek) interpretation of Egyptian monuments.

PART I
THE FRAME IN RELATION TO THE FLAT DISK EARTH

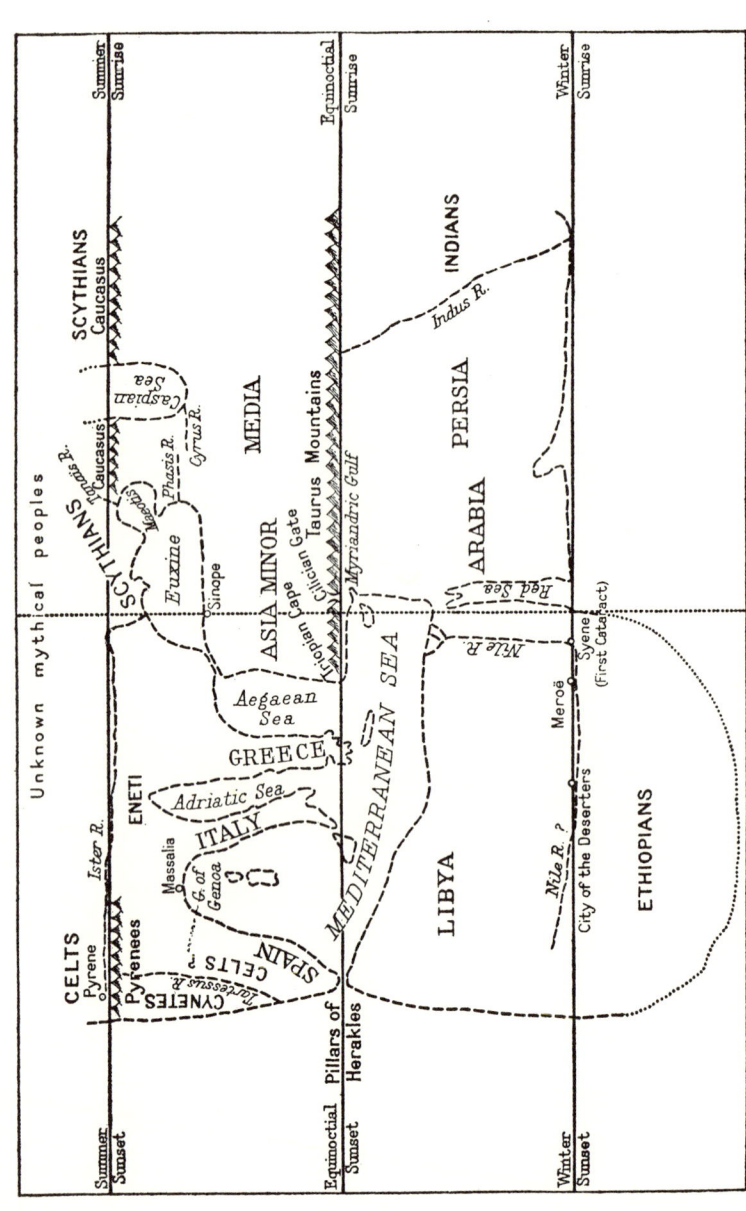

Fig. 1—The probable Greek concept of the *oikumene*. (See foot of opposite page.)

CHAPTER I

THE ORIGINS OF THE FRAME

The earliest conception of the earth entertained by the Greeks was that of a circular disk covered by the " inverted bowl" of the sky. This primitive view, derived directly from the appearance of the horizon and the heavens, was modified by the Ionian philosophers who, in the sixth century before Christ, began speculating in an attempt to reduce appearances to a rational picture. The flat disk was retained, but the heavens were broken up into a series of bands of varying breadth, conceived as circling round the earth all in one plane. The notion was obviously suggested by observations of turbid water made to rotate in a container: as the rotary motion slows down the suspended particles are seen to settle in graded bands, the heaviest collecting at the center and remaining stationary, while the finest continue to be carried round by the circling fluid. So the solid earth, still regarded as the disk suggested by the horizon, was thought to stand still, while the outer bands, composed of mist and fire, continue to revolve about it. This was a very bold hypothesis. While it saved the appearances in respect to the earth, it did away at a stroke with the notion of the " inverted bowl" or hemisphere of the sky. Presumably the appearance of a hemisphere was explained as the effect of distance, as Xenophanes [8] and

---

[8] Aëtius, II, 24, 9 (Aëtius is always cited in the present work according to the reconstruction of his epitome by Hermann Diels, *Doxographi graeci*, Berlin, 1879). The same conception here attributed to Xenophanes underlies the statement regarding Anaximenes in Hippolytus, *Refutatio omnium haeresium*, I, 7, 6 (Hermann Diels, *Die Fragmente der Vorsokratiker*, 5th edit., Vol. 1, Berlin, 1934, p. 92, lines 19-20).

FIG. 1—Sketch map illustrating the probable Greek concept of the general relationships of the *oikumene* to the frame before the time of Eratosthenes and embodying the Persian map of the time of Darius. This is not strictly a reconstruction, since no definite information is available in regard either to the manner in which the details of the coast line appeared on the Greek maps or to the relative distances separating the various features indicated.

8 THE FRAME OF THE ANCIENT GREEK MAPS

Anaximenes in the sixth century explained the apparent circling of the sun. The bands, in which the heavenly bodies were situated, were, however, not in the positions presupposed in the hypothesis of the vortex, for they were obviously not in the same plane as the terrestrial disk. This fact was explained by the assumption of a dip (ἔγκλισις)[9] of the earth to the south, which brought the sun, for example, above the horizon during the day; and to periodic changes in the angle of dip between summer and winter were ascribed the changes of the seasons and also the oscillation of the waters on and under the earth.[10]

Early Astronomico-Geographical Observations

We see at once that the natural philosophers from the first regarded the earth as a part of the cosmos, as the center about which the whole revolves. With the Ionians and those who followed them the celestial phenomena, though exciting great interest, were secondary to the earth in importance. We shall see in the sequel (pp. 83f.) that it was otherwise with the Pythagoreans, who betrayed no interest in geography. From Homeric times certain stars, their risings and settings, were known, but in the *Iliad* and the *Odyssey* and in Hesiod they were of note chiefly as indices of seasons and weather.[11] Hesiod used the summer and winter solstices[12] as points from which to reckon the time for pruning the vine or setting out on a voyage. The *Odyssey*[13] discloses the knowledge that the Bear does not

---

[9] The ἔγκλισις was apparently accepted by all the Ionians.
[10] Presumably Plato (*Phaedo*, 111 c ff.) reflects these old theories.
[11] *Iliad*, V, 5; XI, 62; XVIII, 486f.; XXII, 317f.; XXIII, 226; *Odyssey*, XII, 3; XIII, 93ff.; XXIV, 12.
[12] *Works and Days*, 564ff., 663ff.; see Plato, *Republic*, 527 d.
[13] V, 272ff. Hesiod and Thales were credited with works on nautical astronomy. Callimachus (fr. 94, in Rudolfus Pfeiffer, *Callimachi fragmenta nuper reperta*, Bonn, 1921, pp. 43-44) speaks of Thales in a way that suggests that he measured the declination of the Wain to determine latitude.
(*continued on next page*)

set and that its location in the north could be used as a guide by sailors at night. Little appears to have been added to this store of knowledge when the first general maps were attempted. Astronomical observations of so elementary a character were not adapted for use as the bases of maps. The fact that the Bear does not set, being regarded simply as a fact, did not even suggest an Arctic Circle.

There are, to be sure, several passages in the *Odyssey* that have led to much speculation. One [14] speaks of Aea, Circe's isle, as the " risings of the Sun and the dwelling place of Dawn"; but even if we could locate Aea on a map in consistency with the Homeric account, we should be nothing profited by this statement. Another passage [15] tells of the land of the Laestrygonians, where the paths of night and day lie so close together that a man who could dispense with sleep might there earn a double wage. This has been taken to imply a knowledge of the land of the midnight sun; but it is more likely, perhaps, to be reminiscent of Utopian concepts of a land, like the New Jerusalem, where there would be no night.[16] The reference, however, to the land of the Cimmerians,[17] which was always shrouded in mist and cloud so that the sun never shone upon it and eternal night prevailed, was more probably based on vague reports of the Cimmerians while they still dwelt in their ancient home north of the Pontus. In the same region Herodotus [18] heard of people who slept for six months. A later writer of the school of Democritus, Bion of Abdera, who presumably lived in the fourth cen-

Continuation of footnote 13.
This is of course unhistorical. Actually sailing by night was little practiced in early times. See Homer, *Odyssey*, V, 271-277. On land (see Sophocles, *Oedipus tyrannus*, 795) one might orient oneself by night and follow the direction by day.
[14] *Odyssey*, XII, 3.
[15] *Odyssey*, X, 81ff.
[16] *Revelation*, XXI, 25; XXII, 5. Many ancient peoples thought of the " earthly paradise " as situated in the north.
[17] *Odyssey*, XI, 14ff.
[18] IV, 25.

tury before Christ, stated that there were regions where day and night lasted six months.[19] This was doubtless based on mathematical calculations, but it is hazardous to determine his date by it, as some scholars have attempted to do.

These data, with the possible exception of the last, had no geographical significance for the Greeks, as is obvious in the case of Herodotus. How the suggestions came to them and how they came to be believed are matters for speculation; but there is abundant evidence that from very early times there was communication by various trade routes through unknown intermediaries between Greek lands and the far north, and, besides, notions were current about peoples in the north, like the Hyperboreans, who inhabited a sort of paradise. On the other hand, there were not wanting evidences near at hand and open to Greek traders in the Mediterranean that would have sufficed to suggest by extension speculations of this sort. For from Minoan times there had been communication with Egypt, and from early in the eighth century, at latest, with inner Thrace and Scythia, and the contrasts between Ethiopia and the north were such as to compel attention. Most of these phenomena, however, admitted of satisfactory explanation on the basis of the conception of the earth as a flat disk, which the sun in its daily course approaches nearest at sunrise and sunset, east and west, where the burnt-faced Ethiopians were thought to dwell, and more particularly in winter, when its entire traverse is far to the south.

Meanwhile, beginning with the Milesians of the sixth century and continuing in the fifth and fourth centuries, notable advances were made in astronomy, which, it might

[19] Diogenes Laërtius, IV, 58. Hultsch ("Bion aus Abdera," in *Paulys Real-Encyclopädie* . . . . Vol. 3 (edited by G. Wissowa), Stuttgart, 1899, col. 486) concludes that this statement is the result of calculations based on the determinations of Eudoxus and therefore dates from the fourth century.

be expected, would have exerted a revolutionary influence on geography. It has, in fact, been assumed by most historians of geography that this was the case, and it is clear that it was not left to modern writers to draw this conclusion. Aëtius assumed that Xenophanes in the sixth century was familiar with geographical climes and zones,[20] and Posidonius, a Stoic who flourished about 100 B. C., asserted that Parmenides, early in the fifth century, discovered the terrestrial zones. The statements in question are, however, due to unwarranted inferences, the fact being that almost a century was to pass before geographical consequences were drawn from astronomical discoveries and still another century before they affected the maps.

THE FRAME OF THE EARLIEST MAPS

We know, however, that maps were drawn in the sixth century and multiplied in the course of time, so that by the end of the fifth century they were fairly common. How were they constructed and what was the frame within which they fell? That there was a frame even from the beginning is reasonably sure. Anaximander, who is credited with making the first map, conceived a picture of the cosmos with the earth as its center; and we can hardly imagine him as depicting the earth without somehow recognizing certain lines and limits. Herodotus, as has already been remarked, ridiculed the early maps. "I laugh," he says,[21] "to see how many have ere now drawn maps of the earth, not one of them showing the matter reasonably; for they draw the earth round as if fashioned by compasses, encircled by the river of Ocean, and Asia and Europe of a like bigness." From this it has commonly been inferred that the circular form was the essential and

---

[20] Aëtius, II, 24, 9. This is now acknowledged to be an anachronism.
[21] IV, 36 (translation from A. D. Godley's Greek and English edition, Loeb Classical Library, 4 vols., London and New York, 1921-1924).

most significant characteristic of the early maps. This inference, however, is not justified [22]; it was emphasized by Herodotus because he wished to cast ridicule on the notion of the encircling river. The fact that he added that Europe and Asia were represented as of equal dimensions, however, gives us at least one line of division within the maps. From other sources we know that this line ran roughly east and west, Europe occupying the northern and Asia the southern segment.

Regarding the articulation of the earliest maps there is no direct evidence; but we shall presently discover grounds for reasonable inferences sufficient to enable us to reconstruct them in outline. It was natural that the maps, intended to represent the entire earth, should assume a circular form, since the earth was regarded as a disk; the river Oceanus was, as Herodotus says, an inheritance from the poets. The Greeks commonly used the same word (*gē*) indifferently for the whole earth and for its parts, an ambiguity inevitably leading to confusion. To obviate it later geographers sometimes, but not consistently, distinguished between the "earth" and the "inhabited earth" (*oikumene*). Everything favors the conclusion that this distinction was tacitly drawn by the earliest cartographers, though we cannot say how it was indicated on the maps, except that more details inevitably appeared in the parts devoted to the known lands. Beyond the confines of the *oikumene* lay regions peopled by the imagination, which the geographer could no more ignore than the poet,

---

[22] Aristotle (*Meteorology*, 362 b 12ff.) also ridicules the current maps, saying that they represent the *oikumene* as circular. He must have spoken carelessly. It is quite certain that the continental mass, not to speak of the *oikumene*, was not circular, though the map probably was in the earlier times. One suspects that the map of Anaximander may not have included much eastward, beyond the Phasis; but Hecataeus enlarged it to include India and may have represented it as extending far beyond the Indus, even showing the Ganges. See Hecataeus, fr. 299, in Felix Jacoby, *Die Fragmente der griechischen Historiker*, Vol. 1, Berlin, 1934, p. 38, and my "Suggestion Concerning Plato's Atlantis" in *Proc. Amer. Acad. of Arts and Sci.*, Vol. 68, 1933, pp. 212ff.

THE ORIGINS OF THE FRAME

for a certain romantic interest attached then as now to the " outskirts " of the world. Thus Herodotus says [23] of these remote borderlands that it is reasonable to suppose that, as they enclose and wholly surround all other lands, they should possess the things we deem best and rarest. What is of especial importance is that the " outskirts " (ἐσχατιαί) were recognized, and that they lay, not beyond the limits of the whole earth, but beyond the fringe of the known earth. There were, therefore, acknowledged limits of the known earth, and these limits tended to approximate those of the *oikumene*.

Naturally the limits were not absolutely fixed; the scope of Greek knowledge, however, was not greatly enlarged from the close of the sixth century until the time of Alexander's conquests, and even then the additional geographical information was confined chiefly to details in the East. Nothing is plainer than the fact that the historians of Alexander without exception [24] depended on quite early maps, which were based upon the eastern conquests of Darius; and we may be sure that the map of Hecataeus took account of the information thus gained.[25]

The Greeks were a seafaring people, and the parts of the earth best known to them were the Aegaean and Mediterranean and the bordering lands. For them the Euxine was essentially an extension of this area. By the middle of the sixth century, when the first general maps were attempted, their knowledge of the Mediterranean basin, though by no means accurate or complete, was sufficient to enable them to form a conception of its extent and

---

[23] III, 106, 116; see Theophrastus, *Historia plantarum*, IX, 15, 2.

[24] Deïmachus, who is sometimes wrongly made one of their number, may have used the map of Dicaearchus, as will be pointed out later.

[25] Hecataeus certainly mentioned the voyage of Scylax; and, if he embodied the " Persian map " of Herodotus (IV, 37ff.), as seems equally certain, he must have drawn the southern limit of Asia from India to Arabia as practically a straight east and west line; for Herodotus (III, 107) thought Arabia the southernmost land of the *oikumene*.

14 THE FRAME OF THE ANCIENT GREEK MAPS

general outlines. The longitudinal axis of the Mediterranean naturally provided the median axis of their maps. At first the eastward production of this axis must have depended on the extent to which the oriental lands were known. In the west the Pillars of Herakles marked the boundary of the Greek world; and to north and south knowledge of some sort extended over about equal distances from the median axis.[26]

These limits of geographical knowledge are vaguely indicated by the poets. Pindar, who repeatedly mentions the Pillars as the bourn beyond which man may not venture, names the Phasis and the Nile as the northern and southern borders of the Greek world.[27] More suggestive still are passages in which mythical events are localized; for, though the distinction between myth and fact is not expressly made, there can be no doubt that the poets, following common practice, relegated the fabled peoples to the outskirts of the known world. In this regard nothing is more instructive than the picture presented by Aeschylus in his treatment of the myth of Prometheus. The scene of the *Prometheus Bound* is laid in the Scythian tract,[28] described as an uninhabited desert. Useless as may be an attempt to trace on a chart the course of Io's wanderings in detail, in broad outline her route is clear enough. We are told that the land of the Scythians and Colchians, who dwell about the Maeotis, is the most remote of the earth[29]; thence Io is bidden to travel eastward, avoiding the Chalybes on

---

[26] Reference should be made to the illuminating treatment of early Greek maps by Professor J. L. Myres ("An Attempt to Reconstruct the Maps Used by Herodotus," in *Geogr. Journ.*, Vol. 8, 1896, pp. 605-629). Though it contains statements that may be questioned and that its author may now disapprove, it is of fundamental importance. Unaccountably it has been neglected to the detriment of the history of geography. I can recall only a single continental reference to it, and that by a Frenchman.
[27] *Isthmia*, II, 41f.
[28] 1f. Here οἶμος clearly is equivalent to ἀκτή, a term that came to have a technical sense. See Herodotus, IV, 38.
[29] 415ff.

THE ORIGINS OF THE FRAME 15

her left and reaching Scythia.[30] She will arrive at the Hybristes River,[31] which she may not cross before reaching the very height of Caucasus (the northern limit of the *oikumene*), loftiest of mountains. Thence she must turn southward[32] and suffer herself to be guided by the Amazons, who will lead her to the Cimmerian Strait, where she will pass from Europe into Asia.[33] From that point she must fare eastward again to the abode of the Gorgons, the Phorcides, and the Scythian Arimaspians.[34] So journeying, she will again reach an utmost land, inhabited by a black race (the Indians), who dwell near the springs of the Sun, whence flows the Ethiop River. Following its course, she will arrive at the cataract where the Nile issues from the Bibline Mountains to empty its waters by seven mouths into the sea.

Except for the west, this account gives a sufficient indication of the frame of the *terra cognita*. Though there is some confusion, which may be due in part to dislocation of verses,[35] it is plain that the northern line ran along the borders of the Scythians until it reached India, which lay at the eastern end. From there, the southern line skirted the borders of Ethiopia as far as the first cataract of the

---

[30] 707ff. Aeschylus apparently located the Chalybes on the outskirts of the *oikumene*, north of the roughly eastward course of Io—not in Armenia as was later done.

[31] See Kiessling, " Hypanis," in *Paulys Real-Encyclopädie* . . . , Vol. 9. (edit. by G. Wissowa and W. Kroll), Stuttgart, 1914, col. 210.

[32] This course is obviously a concession to local associations with the myth of the Amazons.

[33] Herodotus (IV, 45) also knew this boundary, though he commonly placed it at the Phasis (IV, 45; II, 103). If, as seems certain, the earliest maps represented the Caucasus as the boundary between Europe and Asia in the east and drew that range of mountains approximately from west to east, beginning at the northeast extremity of the Euxine, either the mouth of the Tanaïs or that of the Phasis might be regarded as on or near the boundary. One wonders whether the reference of Aeschylus to the Cimmerian Strait may not have been a mistake. Could he have thought of the Caspian Gates, which lie between the Caspian Sea and the eastern end of the Caucasus? Here, no doubt, the Scythians entered Media. According to the old maps this pass also would lie in line with the Tanaïs and the Phasis.

[34] 790ff.

[35] Or possibly to conflation of several editions.

16   THE FRAME OF THE ANCIENT GREEK MAPS

Nile. The indicated route of Io from India to Egypt certainly reflects an old map, for even in the days of Alexander it was thought that the Indus was the upper course of the Nile. Beyond these limits all was supposed to be desert [36]—lands uninhabited except by fabulous beings. That these limits were indicated in the map of Hecataeus and his accompanying geographical treatise is practically certain, as will presently be seen. It is obvious that, in the region north of the Euxine, the northern limit of the *oikumene* lay not far from that sea. Regarding the west Aeschylus furnishes us with little information in the *Prometheus Bound*; but from the *Prometheus Unbound* we learn that among the storied peoples whom Herakles was to see were the Ligyes or Ligurians.[37] If we may assume, as I am sure we must, that the Ligurians, who lived along the northern shore of the Gulf of Genoa, were supposed to dwell along the same " parallel " as the Scythians, it follows that Aeschylus must have regarded that gulf as extending far to the north. We shall presently bring forward further evidence that this was the way in which at least some early maps represented the western part of Europe.

That this frame is not merely an inference from the account of a poet, who might take such liberties as he chose, can be shown with absolute certainty. Ephorus, a younger contemporary of Plato, composed a history in

[36] The " deserts " of course mark the limits of the *oikumene* proper. We have seen that the Scythian tract in which Prometheus was chained was desert. In India Herodotus (III, 98; IV, 40) shows deserts; in Ethiopia and Libya likewise. Arrian (*Anabasis*, VI, 1) shows that the Indus had been supposed to lose its identity in a desert and to emerge as the Nile. Herodotus (V, 9-10) thinks of Scythia north of the Ister as desert and uninhabitable because of the cold. As to the fabled inhabitants of these " outskirts " of the *oikumene*, they could be shifted from one point to another almost at will. Thus the Gorgons, whom Aeschylus located in the north, were (in the person of Medusa) supposed to be in Ethiopia (Herodotus, II, 91).

[37] Hesiod (fr. 232, Alois Rzach's edit., Leipzig, 1894; Strabo, VII, 3, 7) mentioned Ligurians with Ethiopians and mare-milking Scythians certainly as inhabitants of the " outskirts " of the earth. Though Spain was sometimes regarded as their habitat, this must have been by extension or by mistake.

THE ORIGINS OF THE FRAME

thirty books, carrying the story down to 340 B. C. Unfortunately his work has not survived to our times, except in fragments. One feature of his treatise was a rather full treatment of geography, in which he depended on Ionian maps. Among the extant fragments is one of exceptional interest, preserved more or less completely by Cosmas, a Christian writer.[38] It consists essentially of an outline of the *oikumene* in the form of a parallelogram,

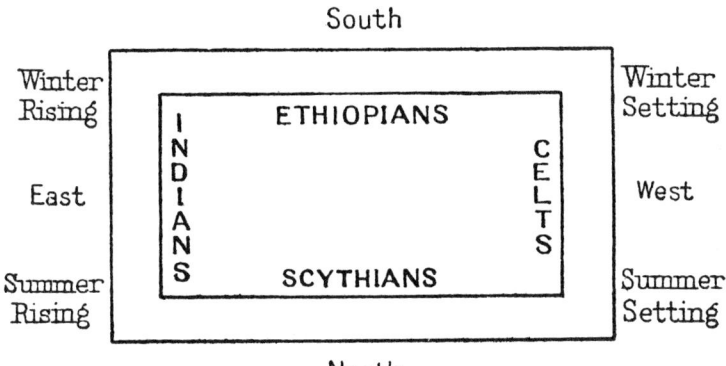

FIG. 2—The parallelogram of Ephorus

each side of which is the boundary of one of the outlying nations. On the west are the Celts, on the north the Scythians, while the Indians lie along the east, and the Ethiopians along the south. It is at once obvious that, except for the Celts, the frame that Ephorus gives is precisely that which one must infer from Aeschylus. It will presently be seen that there is good reason for tracing

---

[38] Fr. 38, in Carl and T. Müller, *Historicorum graecorum fragmenta*, Vol. 1, Paris, 1841, pp. 243-244; fr. 30, in Felix Jacoby, *Die Fragmente der griechischen Historiker*, Vol. 2, Berlin, 1926, pp. 50-51. See also *The Christian Topography of Cosmas Indicopleustes*, edited with geographical notes by E. O. Winstedt, Cambridge, England, 1909, pp. 80-83, 344; *The Christian Topography of Cosmas, an Egyptian Monk*, English translation with notes by J. W. McCrindle, London, Hakluyt Society, 1897, pp. 73-74.

it back at least to Hecataeus of Miletus.[39] Interesting as this diagrammatic [40] outline of the Greek map is in itself, were this all that the fragment gave, it would not contribute much to our understanding of the basis on which it rested.

Help, however, is afforded by other data in the same fragment. Here the northern line is said to run from the summer sunrise to the summer sunset, while the southern runs from the winter sunrise to the winter sunset, and these two sides are declared to be longer than the eastern and western. Regarded by themselves these statements are enigmatic, but they become intelligible and clear when one takes account of the statements of other authors who depended on Ionian maps. The fundamental conception is entirely in agreement with that of the earth as a circular disk, which was suggested by the horizon. The earlier Greeks had no special reason to regard the horizon as shifting with changes in the position of the observer. Hence, to them, the horizon furnished certain fixed points of paramount importance—those where the sun rises and sets at the solstices. Divided by these points and those of the equinoxes, the horizon served in lieu of a compass for designating directions, especially east and west, north and south being more commonly indicated by reference to the winds.[41] Naturally, when a complete wind rose was con-

---

[39] This is certain, except possibly for the Celts. The Indians and the Ethiopians were known to him; and as for the Scythians it is clear that Herodotus was familiar with a geographer, undoubtedly Hecataeus, who called all the northern peoples Scythians. The fragment relating to Caspapyrus seems to show that he called the entire northern belt "the Scythian ἀκτή." See my "Suggestion Concerning Plato's Atlantis," in *Proc. Amer. Acad. of Arts and Sci.*, Vol. 68, 1933, p. 214, and "Hecataeus and the Egyptian Priests in Herodotus, Book II," in *Memoirs Amer. Acad. of Arts and Sci.*, Vol. 18, Part 2, 1935, p. 105, note 98. See also note 79 below.

[40] Professor Myres (*op. cit.*) made abundantly clear the diagrammatic character of the early maps.

[41] I had noted this fact long ago; more recently it has been pointed out by A. Rehm ("Antike Windrosen," in *Sitzungsber. Münchener Akad., Philolog.-Hist.-Klasse*, 1916, No. 3, pp. 27ff). Clear examples are Hippocrates, *De*
(*continued on next page*)

THE ORIGINS OF THE FRAME

structed, account was taken of these points on the horizon; but the scheme of Ephorus and of others who reproduced the Ionian maps was not derived from the wind rose. It was, rather, a definite geographical frame, which can still be located with approximate accuracy, especially as to its north and south sides. The *oikumene* continued to be regarded as elongated, but the ratio of its breadth to its length was differently estimated according to the information available to the different writers.

Greek and Roman writers frequently praise the climate of Asia Minor and of Greece [42] for its moderation and its salutary influence on the inhabitants. This might be regarded as the result of empirical observation, because it is in accord with fact. These lands are actually favored by climatic conditions of an equable yet invigorating character when compared with the extremes of heat and cold to be experienced farther south and north. But we are fortunate enough to possess clear evidence regarding the theoretical explanation of these phenomena. The Hippocratic treatise *On Climate, Waters, and Situations*,[43] which may be dated with reasonable certainty in the latter half of the fifth century, tells us that they are due to the circumstance that "Asia Minor lies midway between the sunrises." This leaves no room for doubt that the sunrises were supposed to have a definite position and that Asia Minor was believed to lie along the equatorial axis of the map; if one were

Continuation of footnote 41.
aëre, etc., chapter 1, and Heraclitus, fr. 120 (Hermann Diels, *Die Fragmente der Vorsokratiker*, 5th edit., Vol. 1, Berlin, 1934, p. 77). Rehm, who has paid little attention to geography, misunderstood or ignored many passages, which will presently be considered, and consequently assumed that mere direction (defined with reference to the wind rose, as to a compass) was meant, when in fact there was a reference to a definite location on the map. The fragment of Heraclitus is our earliest evidence of an attempt to define a meridian.

[42] See Herodotus, I, 142; III, 106; Hippocrates, *De aëre*, etc., chapters 12, 23; Plato, *Timaeus*, 24 c; *Critias*, 108 c, 111 e; Aristotle, *Historia animalium*, 606 b 17ff., 607 a 9f.; *Politics*, 1327 b 20ff.; Pliny, *Naturalis historia*, II, 190.

[43] Chapter 12. There is no doubt that all this comes from the map and the geographical treatise of Hecataeus of Miletus.

inclined to question this conclusion one should note the way in which, in the same treatise, the Scythians and the Egyptians (the Ethiopians are nowhere mentioned [44] in the tract) are contrasted, the former on the north, as subject to extreme cold, the latter on the south, as exposed to extreme heat. Obviously we have here the very frame subsequently adopted by Ephorus; and there can be no doubt that it comes from older Ionian maps. Later writers, accustomed to maps constructed on the principles of mathematical geography, misunderstood this scheme, taking it to refer to our temperate zone [45]; consequently certain corruptions have crept into the text of this treatise and of several others.[46]

We have, then, three clearly indicated parallels on the Ionian maps, corresponding to the tropics and the equator of our maps, but drawn in places where we should not think of locating them. That these lines were relative to the Greek horizon is obvious, and the contrasted extremes of heat and cold, among the Scythians and the Egyptians (Ethiopians), sufficiently indicate the practical coincidence of the frame with the limits of the *oikumene*. Thus far, however, the data that have been considered do not enable us to locate these lines at all adequately. Aeschylus and Ephorus do, indeed, indicate them for the rather vague east, but for the western world we must look elsewhere.

Dependence of Herodotus on the Ionian Maps

At this point Herodotus comes to our aid. Critical as he was of the older geographers in matters that he regarded as speculative, he was nevertheless wholly dependent on their maps. Though he knèw a considerable number, he

---

[44] They are, however, undoubtedly included among the "Libyans," who are repeatedly mentioned.
[45] For Galen, see Hugo Berger, *Geschichte der wissenschaftlichen Erdkunde der Griechen*, 2nd edit. Leipzig, 1903, p. 123, note 3.
[46] Some are discussed farther on, others I hope to take up on a later occasion.

THE ORIGINS OF THE FRAME

nowhere indicates an essential difference between any of them—a fact that certainly suggests that they were all very much alike. As he wrote in the latter part of the fifth century, we should be certain that his maps were of Ionian origin even if he had not specifically criticized the geographical notions of the " Ionians " in contexts that leave no doubt that his references are specifically to Hecataeus of Miletus.[47] In regard to the " tropics " he gives us a valuable hint in his discussion of the causes of the Nile floods.[48] We need not rehearse his theory in detail: it amounts essentially to this, that we should not speak of a flood of the Nile at all, since what is commonly so regarded is only the normal flow of the river, which is greatly reduced in the winter. At that time, he says, the sun, traversing inner Libya (Africa), stands over the Nile and consequently, aided by hot winds, produces vast evaporation. In the summer, however, the sun turns northward and draws equally from all rivers because it approaches midheaven. If the seasons were reversed, the sun would traverse the inner part of Europe, as it now passes over Libya (Africa), and would affect the Ister (Danube) as it now affects the Nile.

From this statement it is evident that to Herodotus the Nile and the Ister corresponded, lying symmetrically about the median line that passes through Greek lands and may be called the " equator " of the Ionian map. Obviously the Nile was regarded as on or near the winter " tropic ": consequently, the Ister must be located along the summer " tropic." Though not expressly stated, this is certainly implied in the description of the course of the sun, which

[47] See Felix Jacoby, " Hekataios," in *Paulys Real-Encyclopädie* . . . Vol. 7 (edited by Georg Wissowa and Wilhelm Kroll), 1912, cols. 2666ff.

[48] II, 24-27. Pindar (*Olympica*, III, 14-18, and *Isthmia*, VI, 23) shows that the Nile and the Ister were coupled as a pair, limiting the *oikumene* on the north and the south. He locates the fabled Hyperboreans beyond the Ister.

in summer traverses inner Europe as it traverses inner Africa in winter. Herodotus [49] knew of maps that divided the *oikumene* in northern and southern halves of equal size, Europe being to the north, Asia (with Libya) to the south. Of course, he does not say—as a man of common sense, he could not say—that in summer the sun at noon stands directly over the Ister.[50] This may be to his credit, if one chooses to think so, but it shows that, like everyone else in his time, he did not regard the meridian height of the sun as significant. It was the Greek horizon, with the points on it marking the risings and settings of the sun, that really mattered.

HERODOTUS ON THE NILE AND THE ISTER

What Herodotus says in connection with the Nile floods needs to be supplemented by other data, if we are to trace the frame of the Ionian maps. Here again the views of Herodotus regarding the Nile and the Ister prove to be helpful. In considering the source of the former, he tells us that the course of the river is known beyond the limits of Egypt for a distance of four months' travel,[51] but he gives no clear indication of its direction immediately above the first cataract,[52] though it is perhaps significant that in his account of the expedition of Cambyses to the land of the Long-lived Ethiopians, which was evidently believed to lie to the south, there is no indication that the expedition

---

[49] IV, 36.
[50] The expression "the middle of the heaven" obviously does not mean the zenith, though we shall later discover an error of $12\frac{1}{2}°$ in a supposed observation of the zenith (see below, note 261). Hippocrates (*De aëre*, 19) says that at the end of its northward progress the sun "approaches most nearly" to the Scythians.
[51] II, 31.
[52] In view of the fact that Ethiopia is the border nation of the *oikumene* on the south it is of interest to see precisely where Herodotus thought the border lay. From II, 29, it would appear that it was at or near Tachompso, 12 *schoeni* above the first cataract.

THE ORIGINS OF THE FRAME

followed or paralleled the Nile.[53] At one point, however, in summing up the total known length of the Nile to the City of the Deserters, Herodotus states that its course runs from west to east[54]; this datum, we may assume, was derived from a map. Herodotus then goes on to say that beyond the City of the Deserters no one has definite knowledge of the river, because the land is a desert by reason of the heat. However, certain men of Cyrene, who claimed to have gone to the oracle of Ammon, told him a tale that had been related to them by Etearchus, king of the Ammonians, who in turn had learned it from Nasamonians who had visited him. The Nasamonians inhabited the region extending west from Cyrene to the vicinity of Carthage. It appears that certain young men of that tribe, wishing to explore the desert beyond the limits of ordinary travel, first went inland beyond the inhabited coastal tract and then penetrated far beyond the next tract, which was the home of wild beasts. After many days of traveling (south?) westward, they came upon a large river flowing from west to east, in which there were crocodiles. From this account Herodotus and Etearchus (and as the story implies, the Cyreneans) inferred that the river in question was the Nile. Herodotus himself concluded that the in-

[53] III, 17-25. Herodotus (III, 114) says that Ethiopia, the utmost inhabited land, extends from the south, where it turns, to the sunset. The phrase is certainly curious. Apparently in saying that the south ("midday") turns to the sunset Herodotus was thinking of the sun, after noon, declining from its highest station; but he meant also that Ethiopia extended westward from "true south." ("True south" was probably regarded as a point on the fixed horizon, just as the extremities of the median longitudinal axis of the earth may have been regarded as "true east" and "true west"; cf. Strabo, I, 2, 25-28, and Heraclitus, fr. 120, Diels' edit.) This seems to locate Ethiopia from "true south" to the extreme southwest. I agree with Heinrich Stein, in his note on this passage (*Herodotos*, Berlin, 1877, Vol. 2, p. 125), in thinking that Herodotus referred to the land of the Long-lived Ethiopians. If this is correct we should expect their capital to lie in the southwest of Africa; and the fact that there is no reference to the Nile in connection with the expedition suggests that Herodotus conceived the march as directed to a region south of the west-east course of the upper Nile, as described below.

[54] II, 31.

ference was reasonable. "For," says he,[55] "the Nile flows from Libya, and right through the midst of that country; and as I guess, concluding from the known to the unknown, it takes its rise from the same degree of longitude as the Ister. The latter river flows from the land of the Celts and the city of Pyrene through the very midst of Europe; now the Celts dwell beyond the Pillars of Herakles, being neighbors to the Cynesii, who are the westernmost of all the nations inhabiting Europe. The Ister, then, flowing through the entire length of Europe empties into the Euxine at the point where lies the Milesian colony of Istria." In another passage [56] Herodotus, repeating his assertion that the Ister, rising among the Celts—after the Cynetes (or Cynesii) the westernmost nation of Europe—flows across the entire length of Europe, says that it empties into the Euxine on the borders of Scythia. The fact that the Ister is here represented as flanking Scythia shows that, as in Pindar, it at least approximates, if it does not mark, the northern limit of the *oikumene*.

There can be no doubt that Herodotus, in sketching the course of the Ister, was following the indications of a map. This is made as clear as possible by several points that should be emphasized. Indeed, the formal way in which he twice describes its course, in almost identical terms, must suggest this; for he had personal knowledge of the river only at its mouth. Furthermore, he says [57] that, as it flows through inhabited country, its course is known to many. It need hardly be remarked that the course that he describes could not have been known to many or, in fact, to anyone, because it is neither true nor possible. Yet Herodotus regarded it as true and known, as he distinctly implies in adding that, " concluding from the

---

[55] II, 33 (translation adapted from A. D. Godley's Greek and English edition, Loeb Classical Library, 4 vols., London and New York, 1921-1924).
[56] IV, 49.
[57] III, 34.

THE ORIGINS OF THE FRAME

known to the unknown," the source of the Nile must lie on the same meridian as that of the Ister. He does not, of course, use the term "meridian," though one does no violence whatever to his meaning in so translating his words. In the same context, obviously with a map in mind, if not in hand, he describes [58] a meridian conceived as running straight from the mouth of the Ister via Sinope and the Cilician Gates to the mouth of the Nile. Moreover, to make the courses of the Nile and the Ister as nearly as possible similar, he elsewhere [59] tells us that the Ister, in the neighborhood of the Euxine, flows southeastward. Since he begins the description of the meridian with the Pelusiac mouth of the Nile, which actually flows northeastward, the courses of the two rivers exactly correspond.

We do not know whether Herodotus derived the story of the expedition of the Nasamonian youths from an earlier geographer or had it directly from men of Cyrene, either in their city or in Egypt. It is uncertain also whether he depended on a map for the supposed course of the Nile above the City of the Deserters. However, since he asserts as unquestioned that at that point (and presumably all the way thence to the vicinity of the first cataract) the river flows from west to east and that it divides Libya longitudinally into equal parts, we may be sure that, in describing the course of the Nile below the City of the Deserters, he was following a map. What he gives as his own conjecture relates solely to the headwaters of the river above that city. At all events Herodotus affords evidence sufficient to determine in a general way the location of the line that marked the winter tropic of the Ionian map.

[58] III, 34.
[59] IV, 99.

CHAPTER II

THE SOUTHERN LIMIT OF THE FRAME

INDIA AND ETHIOPIA

The evidence thus far cited suffices to indicate in a general way the location of the Ionian tropics, which formed the limits of the *oikumene* and the frame of the Ionian map on the north and on the south; but the picture may be made more complete by taking up a number of particular questions. It will be well to follow the lines laid down by Ephorus and determine as far as possible where they lay on the map. Beginning with the southern tropic, because it involves fewer problems, we observe that its eastern end was placed at the southern extremity of India, a country that extended from the summer to the winter sunrise. The southern limit of India was located at the winter sunrise even by as late a writer as Deïmachus early in the third century, and it is curious that the statement was not challenged by Eratosthenes or by Strabo,[60] who reports it.

There is, as has already been said and as will later appear, abundant evidence that the historians of Alexander used old Ionian maps. That being the case, it is not improbable that Alexander himself had such maps when he marched to India. Now, Arrian[61] tells us that, when

---

[60] II, 1, 19. Apparently not only Strabo but Eratosthenes also criticized Deïmachus for referring to the *autumnal* equinox as the northern boundary of India, on the ground that there is no difference between the autumnal and vernal equinoxes (see pp. 48f., below). If this be true, it is probable that Eratosthenes understood, as Strabo apparently did not, the familiar use of the terms by the Ionian geographers.

[61] *Anabasis*, VI, 1, 2f.

THE SOUTHERN LIMIT OF THE FRAME 27

Alexander found, in a river in India, crocodiles and a certain bean, both of which were known to him only from the Nile, he concluded that the Indus was the source of the latter river and that the Indus in flowing through desert lands lost its proper name and among the Ethiopians and Egyptians came to be known as the Nile. Although Arrian does not mention maps in this connection, one naturally asks whether Alexander did not derive the suggestion from a map. Even if this conclusion cannot be established, there is good reason to suspect that it is true. First of all, the notion that the Indus disappeared from sight in a desert is worth noting, because the *oikumene* was regarded as entirely surrounded by desert lands. Furthermore, the wanderings of Io, as we have seen, led her from India (apparently overland) to the cataracts of the Nile, in a course that was evidently conceived as conducting her along the borders of *terra cognita*. As late as Ptolemy there were still those who believed in a land bridge connecting India with Africa. Assuming, then, that the notion of Alexander, which he later learned to be unfounded, was suggested by a map, we have every reason to regard his map as old, or based on an old one—an older map, indeed, than that of Hecataeus, since the latter accepted as true the voyage of Scylax down the Indus to its mouth, which made impossible the identification of the Indus and the Nile.[62]

Such a map, we may be sure, must have suggested a course of the Indus-Nile, or at least a tract of desert beyond the closed Erythraean Sea, running approximately from the southern limit of India to the cataracts south of Egypt. If this tract bore a name on the map, the name was "land of the Ethiopians." The Homeric statement [63] that the Ethiopians were divided in twain, some

---

[62] It is not certain, however, that Hecataeus, following Euthymenes, sought the source of the latter river in the west.
[63] *Odyssey*, I, 22-23.

dwelling by the rising of the sun, others by its setting, would appear to reflect, if it did not give rise to, the same notion that Ethiopia extended along the entire southern limit of the *oikumene*.

The Ethiopians, however, who were always regarded as the border people in the south, were best known to the Greeks as dwelling beyond the southern limit of Egypt, which was then, as now, at the first cataract.[64] Herodotus, as we have seen, did not believe in an Indus-Nile flowing from the east but sought the source of the Nile in the west. Whether this suggestion came from Euthymenes through Hecataeus, as has been asserted, is open to question. There is, however, no doubt that the upper course of the river, in his view, lay along the winter tropic, and the evidence for the identity of the river discovered by the Nasamonian youths is the same as that which prompted the suggestion of Alexander: a river with crocodiles must be the Nile.

Africa

If, now, we seek the location of the southern tropic in Africa, we have several bits of evidence. Attention has already been called to the indications of the text of Herodotus pointing to the conclusion that, in his opinion, the bend of the Nile from an eastward to a northward course was at or near the first cataract. Several other data suggesting the same conclusion may be adduced. Thus, after relating the story of the circumnavigation of Africa by the Phoenicians in the time of Necho, Herodotus says [65] that it proved that Africa was of no very great size. Since he knew approximately the length of the northern coast from the Isthmus of Suez to the Pillars, he can only have been referring to its southward extension. Again it is

---

[64] See note 52, above.
[65] IV, 42. Here Africa is said to have a breadth from north to south far less than Europe.

significant that, having an approximate knowledge of the length of the Red Sea, he declares Arabia to be the southernmost inhabited land,[66] thus placing its southern limit about on a parallel with southern India. Now, since southern Arabia gives the southern limit of the *oikumene*, which according to the Ionian scheme was on or near the winter tropic, marked in Africa by the west-east course of the Nile, and since the upper course of the Nile was thought to divide Africa into equal parts, we may be sure that the supposed bend of the river lay between the first cataract and Meroë, probably quite close to the former. The continent, therefore, could not have been conceived as very large as compared with Europe and Asia. Furthermore, since the *oikumene* was on all sides bounded by deserts, there was no reason for placing its southern limit very far inland, especially as the Ethiopians, the border people, were known to live along the borders of Egypt. Indeed, the last people toward the desert, Herodotus says, were those who dwelt along the "brow," or ridge of sand that he believed to extend from Thebes westward to the Pillars of Herakles.[67] He can hardly have intended the statement that the ridge ended at the Pillars to be taken literally. He must have meant that it extended as far westward as the Pillars, though lying considerably farther south, since he knew of inhabitants along the northern coast of Africa and in the story of the Nasamonians he had heard of pygmies, who were reached only after a considerable strip of desert was crossed. These pygmies were, of course, Ethiopians.

[66] III, 107.
[67] Herodotus, IV, 181-185. This "brow" lies southward of the inhabited lands bordering the Mediterranean and the belt infested with wild beasts. Along this ridge are oases, the first being that of the Ammonians. Farther on there is the land of the Lotus Eaters and other fabulous tribes, such as the Garamantes, who hunt the cave-dwelling Ethiopians; the last-named must therefore dwell not far away. The Atarantes, who follow next in order, find the sun so hot that they curse and revile it. Beyond this ridge southwards Libya is sheer waterless desert, where even wild beasts are not to be found. Obviously we have reached the limit of the *oikumene* in this quarter.

30 THE FRAME OF THE ANCIENT GREEK MAPS

All these data give a harmonious picture in keeping with the frame of Ephorus, according to whom it was reported by the Tartessians that the Ethiopians extended westward to the sunset.[68] No doubt this datum, derived from the Tartessians, accounts for the western terminus at the winter sunset of the southern line of Ephorus' parallelogram.

[68] Strabo, I, 2, 26. There is no good reason for changing the reading of the manuscripts, μεχρὶ δύσεως (to the sunset), as suggested in the edition of Strabo of Carl Müller (Vol. 2, Paris, 1858, p. 942), unless to add χειμαρινῆς (winter). See Herodotus, III, 114, and note 53, above.

CHAPTER III

THE NORTHERN LIMIT OF THE FRAME

EAST OF THE ISTER

Having traced the southern tropic, we may now turn to the northern. Certain stretches of this line have already been described. Beginning at the west it follows the course of the Ister until that river bends southeastward to enter the Euxine. From there the line runs eastward along the borders of Scythia to the Caucasus. Aristotle,[69] herein following Ionian maps, says that the Caucasus is the highest mountain toward the summer sunrise. This has carelessly been taken as meaning merely that it lies to the northeast. If that were true, one might ask, northeast of where? Obviously the old maps [70] represented the Caucasus as running approximately due east to the Caspian. It is deserving of mention that this range was the southern limit of the Scythians and the northern boundary of the Persian Empire. Herodotus so informs us and says that the Scythians once passed beyond this limit into the kingdom of the Medes.[71] Where the Caucasus was supposed to lie is a somewhat difficult question. As we have seen, Herodotus knew of old maps that divided the *oikumene* into approximately equal parts, Europe and Asia. But he does not tell us in just what direction the dividing line ran. It can hardly have been regarded as running due east and

---
[69] *Meteorology*, 350 a 28f.
[70] Herodotus, I, 103-104, 203.
[71] III, 97; IV, 12.

west, because at the point where the continents meet along the Dardanelles and the Bosporus the direction is plainly northeast-southwest. Furthermore, the part of the Caucasus first known to the Greeks lies northeast of the Euxine, which was known to lie north of Asia Minor. The two places where the further demarcation was fixed—the Tanaïs (Don) and the Phasis—lie respectively north and south of this range; but the earliest maps seem to have placed the southern end of the Caspian, where the Caucasus proper ends, uniformly too far north. How the Phasis and the Tanaïs were drawn on the older Greek maps is difficult to decide. It seems probable that the course of the Phasis was inferred from an ancient trade route that followed the valleys of the Cyrus and the Phasis, thus connecting the Caspian and the Euxine. It was then assumed that the two rivers were one, and this was called the Phasis. Since the Caspian was at an early date thought to be a bay of the outer ocean, early Argonautic tales conceived of the Argonauts as passing from Colchis through the Phasis to the ocean, whence they returned into the Mediterranean. Herodotus knew that the Caspian was closed at the north, but later the old conception was revived and was accepted by Eratosthenes. The course of the Tanaïs was differently imagined, one view placing its source in the east, so that the historians of Alexander thought the Jaxartes flowed into the Tanaïs. To whom the variant conceptions are to be attributed we have no means of deciding. It is probable, however, that the Tanaïs, when first accepted as dividing Europe and Asia, was thought to rise in the east and to follow a course roughly parallel to the Phasis. When the Tanaïs was later assumed to flow from north to south, the question of the line of division between the continents became acute. No wonder that Herodotus thought it was

THE NORTHERN LIMIT OF THE FRAME

arbitrary.[72] The northeastern part in any case long remained an unexplored and unknown No Man's Land.[73]

How the Caucasus was placed on the early maps is best shown by the historians of Alexander, who gave that name to the entire series of mountains extending across Asia, beginning with the Caucasus proper, continuing in the Hindu Kush, and ending in the lofty chain forming the northern boundary of India. The emphasis laid by them on the fact that Alexander crossed this chain is clearly due to a desire to represent him as conqueror of the entire *oikumene*, whose northern limits he here reached, just as in India he reached the eastern limits and, by conquering Egypt and penetrating the borders of Ethiopia, he reached the limits on the south. He had only to follow the same lines westward to complete the conquest of the entire world known to the Greeks.[74]

Scythians dwelt along the entire northern border according to Ephorus, who herein clearly followed old Ionian maps. Herodotus, indeed, does not concur with this view, tending to restrict the name of Scythian to nations farther west; but occasionally he betrays the fact that others held it. Thus, referring to the tribe called the Black-Coats, whom Hecataeus had mentioned,[75] he says,[76] " they are a different people, and not Scythians." Similarly in other

---

[72] Herodotus, IV, 45. The questions connected with the Phasis and the Tanaïs have been much debated. See Felix Jacoby, " Hekataios," in *Paulys Real-Encyclopädie* . . . , new ed., Vol. 7 (edited by Georg Wissowa and Wilhelm Kroll), Stuttgart, 1912, cols. 2704-2705; and Herrmann, " Tanais," *ibid.*, *Zweite Reihe*, Vol. 4 (edited by W. Kroll and K. Mittelhaus), Stuttgart, 1932, cols. 2162-2166. The latter assumes a Phoenician chart of the tenth century before Christ, reflected in the Old Testament and in the *Book of Jubilees*, which made the Tanaïs the boundary between Europe and Asia. One is here always dependent on unverifiable assumptions. See above, note 33.

[73] See J. L. Myres, "An Attempt to Reconstruct the Maps Used by Herodotus," in *Geogr. Journ.*, Vol. 8, 1896, pp. 605-629.

[74] In his first campaign he had conquered Thrace and reached the Ister (Plutarch, *Alexander*, 11).

[75] Fr. 154 (Carl and T. Müller, *Historicorum graecorum fragmenta*, Vol. 1, Paris, 1841, p. 10).

[76] IV, 20.

instances.[77] North of the Himalayas, which formed the northern boundary of India and in the frame of Ephorus lay on the summer tropic, Strabo locates the Sacae, whom, following Choerilus, a contemporary of Aeschylus, he calls a Scythian nation.[78] Most significant of all, however, is a fragment of Hecataeus[79] referring to Caspapyrus, the Caspatyrus of Herodotus, which according to Stephanus of Byzantium he locates in the "Scythian tract." This city, as is well known,[80] was the point from which Scylax of Caryanda set out, under orders from Darius, to follow the course of "the Indian River"; it is therefore of interest to note that Herodotus says: "Other Indians are neighbors to the city Caspatyrus and the Pactyic country, dwelling northward of the remainder of the Indians." This certainly suggests that they were north of the Himalayas. He adds that their manner of life was similar to that of the Bactrians, whom Strabo placed on a line with the Sacae.[81] Since Herodotus knew the far east only from the map or from the geographical treatise of Hecataeus we have every reason to assume that the map of Hecataeus essentially agreed with the parallelogram of Ephorus along the line of the summer tropic to the east of the Euxine.

The Course of the Ister

The north shore of the Euxine was of course the Scythian land *par excellence*; but the entire northern tract was evidently in the earlier time supposed to be inhabited by Scythians, perhaps because the name was taken as

---

[77] I, 201; IV, 18-21, 99-104.
[78] VII, 3, 9; see Diodorus, II, 35. For the parallel of Eratosthenes, see Strabo, I, 1, 4; II, 14, 18.
[79] Fr. 295 (in Felix Jacoby, *Die Fragmente der griechischen Historiker*): Κασπάπυρος πόλις Γανδαρική, Σκυθῶν δὲ ἀκτή. Here, Jacoby, adopting a conjecture of Sieglin, reads ἀντίη for ἀκτή, and supplies κεῖται. I see no reason to follow his suggestion and prefer to read simply ἀκτῇ. See above, note 39.
[80] Herodotus, IV, 44.
[81] See above, note 78.

THE NORTHERN LIMIT OF THE FRAME 35

synonymous with "barbarians."[82] Herodotus makes the Ister the boundary between Thracians and Scythians, though a few of the latter had penetrated south of the river.[83] We have already noted the general course of the Ister as conceived by him. It was supposed to take its rise far in the west among the Celts. Where these latter dwelt remains to be determined. As the actual source of the river Herodotus names the city of Pyrene. A city of that name is unknown, and a city, moreover, is not especially appropriate as the source of a river. Aristotle, who gives an account of the Ister otherwise closely parallel to that of Herodotus, makes the river rise in the Pyrenees, which is intrinsically more reasonable and fits better into the scheme, since the Celts were always associated with this range of mountains. One suspects that both authors were referring to maps on which the name Pyrene occurred and that Herodotus carelessly took it for the name of a city.[84]

We have then to inquire where the old maps placed the Celts and the headwaters of the Ister. As the river flowed from the west toward Thrace, there is good reason to suppose that its course was believed to lie not very far from the head of the Adriatic.[85] How old the notion was that the Ister divided, issuing by one mouth into the Euxine, by another into the Adriatic,[86] we cannot say; but that Theopompus (378-320 ? B. C.) shared in this belief [87] suggests that it was not of late origin, and it was probably adopted in some early version of the myth of the Argonautic expedition, which was certainly supposed to follow the limits of the *oikumene*. In any case the notion is

---

[82] See Hugo Berger, *Geschichte der wissenschaftlichen Erdkunde der Griechen*, 2nd edit., Leipzig, 1903, p. 366, notes 4 and 5; Polybius, IX, 34, 1.
[83] Herodotus, V, 9-10. Pindar (*Olympica*, III, 13-15) says that Herakles brought the olive of which the victor's crown at Olympia was made from the springs of the Ister, obtaining it from the Hyperboreans. See Arrian, *Anabasis*, I, 3, 1; Strabo, I, 2, 27; Polybius, III, 37, 9. In part the northern peoples were, especially in later times, called Celts.
[84] This instance is not unique.
[85] Herodotus, I, 193; IV, 47, 89, 99; V, 9-10.
[86] Apollonius Rhodius, *Argonautica*, IV, 282, 307ff.
[87] Strabo, VII, 5, 9.

compatible with a location of the river near the head of the Adriatic, and we naturally get the same impression from the statement of Herodotus [88] that the only people known to him along that reach of the Ister were the Sigynnae, whose territory approached the boundary of the Eneti, who lived along the northern end of the Adriatic. When one looks at a map this location seems fantastic, because the head of the Adriatic is not only far south of the course of the Ister in that longitude but is also quite out of line with the Caucasus and the north shore of the Euxine. This, however, signifies little; the Adriatic was in fact not well known until quite late times. It is probable that the early geographers derived their conception of this gulf from the Phocaeans, whose explorations may be dated in the sixth or seventh centuries. There is, moreover, good reason to believe that the Adriatic was supposed to run due north and south, if indeed it was not actually drawn northeast-southwest, making the Balkan peninsula quite narrow at the north.[89] That would place its recesses considerably farther north, even if its length were correctly estimated, which is questionable; for the statement of Dicaearchus [90] that the distance from the Peloponnesus to the head of the Adriatic was greater than that to the Pillars of Herakles is hardly intelligible unless the gulf was thought to be far deeper than it actually is.

The Headwaters of the Ister

How we should plot the imagined course of the Ister farther west is not so easily determined. As we have seen, the river was supposed to cut Europe in two, and of course it lay at or near the northern limit of the *oikumene*. Evidence for this region is scanty. Herakles, as has been

[88] V, 9.
[89] See Felix Jacoby, *Die Fragmente der griechischen Historiker*, Vol. 1, Berlin, 1934, pp. 338f., on Hecataeus, fr. 90-96.
[90] Strabo, II, 4, 2.

THE NORTHERN LIMIT OF THE FRAME 37

pointed out,[91] in the *Prometheus Unbound* of Aeschylus came to the Ligurians, who dwelt along the inner recesses of the Gulf of Genoa. This suggests that in that region the limit of the *oikumene* was not far inland. Berger calls attention to indications that suggest that the Gulf of Genoa also was supposed to extend much farther north than it actually does,[92] which would probably result from drawing the coasts of Italy and Spain approximately north and south.

At this point one faces a difficulty of the first magnitude; for the Ister must cross the course of the Rhone, one of the greatest rivers of Europe. That the Ister was thought to rise west of the actual course of the Rhone admits of no reasonable doubt, because its source was assumed to be in the Pyrenees among the Celts who, after the Cynetes, or Cynesians, were the westernmost inhabitants of Europe. How such a view could be entertained is indeed difficult to understand; certainly it seems inconceivable that it should have originated after Massalia was founded by the Phocaeans (*ca.* 600 B. C.), because the colonists must soon have established relations with the interior, which would disclose the course of the Rhone for a considerable distance from its mouth. It must be admitted that this problem cannot be solved with certainty.

There is, however, a possibility that calls for consideration. It seems to be certain that the western Mediterranean was first made known to the Greeks by Phocaeans and Samians, who maintained friendly relations with each other and also with the Phoenicians. Their first expeditions must have been made not later than the first quarter of the seventh century,[93] and it is stated that the Phocaeans

---

[91] See above, p. 16.
[92] Berger, *op. cit.*, 105f. This is contrary, however, to the supposition of Myres. Much would depend on the point from which the position of the Pyrenees was determined, and there were undoubtedly a number of sketch maps for different sea routes. The earliest information regarding the Pyrenees undoubtedly came from Tartessus and consequently related to the Atlantic end.
[93] Herodotus, IV, 150ff.

won the confidence and friendship of Arganthonius, king of the Tartessians, who would have had them abandon their city and dwell in his land; he gave them the means wherewith to build a strong wall when the power of the Medes grew menacing.[94] The reports of the extraordinary length of this king's life and reign suggest that there may have been several kings of that name.[95] While part of this tale clearly refers to the time after the fall of Croesus (546 B. C.), because the Medes could not have threatened Phocaea before, the relations with Tartessus probably began much earlier. Herodotus says that the Phocaeans were the first to make long sea voyages and that it was they who discovered the Adriatic, Tyrrhenia, and Iberia.[96] This would, in fact, suggest the eighth rather than the seventh century. Massalia was founded later (*ca.* 600 B. C.) but the relations between the Greeks and the Phoenicians and Tyrrhenians must even then have been less strained than they became at the close of the sixth century, when Alalia was destroyed (*ca.* 535 B. C.) as a lesson to the Greeks, a lesson to be further impressed by the invasion of Sicily in force during the Persian Wars.[97] This state of affairs doubtless accounts in part for the relative ignorance of the Greeks regarding the geography of the western Mediterranean basin. From the fragments of his geographical treatise one gathers that Hecataeus knew more about the region than any of his successors before the third century, but even he could not obtain full or accurate information. His presumable visit to those parts must have occurred in the period between the fall of Alalia and the Persian Wars,

---

[94] Herodotus, I, 163, 165. On this and on all questions relating to Tartessus, see the studies of Schulten, a convenient summary of which he has given in his article " Tartessos " in *Paulys Real-Encyclopädie* . . . , *Zweite Reihe*, Vol. 4 (edited by W. Kroll and K. Mittelhaus), Stuttgart, 1932, cols. 2446-2451. His views differ at many points from mine. While I owe much instruction to him, I regard much that he says as too speculative to be accepted.
[95] See Karl Müllenhoff, *Deutsche Altertumskunde*, Berlin, 1887-1900, Vol. 1, pp. 109ff.
[96] I, 163.
[97] Cf. Müllenhoff, *op. cit.*, Vol. 1, pp. 96f.

THE NORTHERN LIMIT OF THE FRAME

when the relations with the Carthaginians were most strained. Moreover, it is doubtful whether he visited Massalia or obtained any information from its inhabitants; for Massalia was a colony of the Phocaeans, who maintained close commercial relations with the Samians, the aggressive rivals of the Milesians. As a Milesian, Hecataeus, who doubtless combined trade with exploration, would hardly have been welcomed in that city. What he learned about Gaul, one suspects, he learned nearer home, possibly through Phocaeans who had returned and who knew only the older traditions that came by way of Tartessus.[98]

This conclusion is suggested by the statement of Aristotle[99]: " From Pyrene (a mountain toward the equinoctial sunset in Celtice) flow the Ister and the Tartessus—the latter beyond the Pillars, the Ister throughout all Europe into the Euxine." It is noteworthy that later one hears little of Tartessus, commonly identified with Gades,[100] on the Guadalquivir. This identification is very doubtful for many reasons.[101] Tartessus is unquestionably the biblical Tarshish. Why it disappeared is uncertain. It was of course on the Atlantic, on or near the Tartessus River, which Aristotle found traced on his map as rising in the Pyrenees. If, as we should naturally infer, the map was based on information derived from Tartessus, the river in question can hardly have been the Guadalquivir; for the source and course of the Guadalquivir lie too far to the

---

[98] Müllenhoff (*op. cit.*, Vol. I, pp. 110f.) maintains that Hecataeus knew neither Celts in Gaul nor Carthaginians in Spain. Certainly the fragments make no mention of the latter. Tartessus, which is mentioned, was founded from the homeland long before Carthaginians entered Spain. Whether Hecataeus knew Celts in Gaul is not certain; the mention of " Celtice " in frr. 54-56 (Jacoby, *op. cit.*) may be due to Stephanus; but from whom, if not from Hecataeus, did Herodotus derive his datum regarding the source of the Ister?
[99] *Meteorology*, 350 a 36ff.
[100] Avienus, *Descriptio orbis terrae*, v. 610ff.
[101] Pseudo-Scymnus (v. 161-164) locates Tartessus two days' sail from Gades. Avienus (478ff.) is very indefinite: Tartessus lies between the " Hispanus ager " about Calpe and the Cempsi, who inhabit the slopes of the Pyrenees.

south. The position given to the Tartessus suggests a river much farther north, such as the Douro or the Tagus. We should, accordingly, seek the site of the Phoenician colony of Tartessus, which for reasons unknown yielded before the Carthaginian Gades, elsewhere—perhaps in the neighborhood of Lisbon or Porto. In any case the connection of the Tartessus with the Pyrenees makes it likely that the position given to the latter was determined by data derived from the Atlantic side[102] and not from the region of Massalia.

In the seventh or sixth century Celts would probably have been known only, or chiefly, north of the Pyrenees.[103] The only Celtic localities mentioned in the fragments of Hecataeus are in Gaul. The source of the Ister was laid by Herodotus, as we have seen, in the far west of Europe, in Celtice, separated from the ocean only by the Cynetes or Cynesians. Why its source was thought to be there we cannot be sure; but a plausible suggestion occurred to Berger,[104] who called attention to the Oestrymnii, a people mentioned by Avienus as living along the Atlantic coast from the Bay of Biscay to Brittany. Their name might suggest the Ister, whose source was known to be somewhere in the west, and their location well suits the conditions. The source of the Ister would thus be placed north of the Pyrenees, and the presumable source of the information would likewise be located on the Atlantic side.[105] We know that Tartessus was interested in the tin trade and had commerce with western Europe at least as far as Albion. On the assumption that Berger's conjecture is right, we must

---

[102] See Müllenhoff, *op. cit.*, Vol. 1, p. 97.
[103] *Ibid.*, Vol. 1, p. 108.
[104] Berger, *op. cit.*, p. 235; see, however, Müllenhoff, *op. cit.*, Vol. 1, pp. 85f.
[105] This conclusion is suggested also by Herodotus, V, 33, where the Celts are said to live "outside the Pillars of Herakles." Just where he thought of the Cynesians as dwelling is not clear. In later times they were located in the southwest of Spain; but there is evidence likewise that they were sometimes located in the northeast corner of Spain, near Narbonne.

THE NORTHERN LIMIT OF THE FRAME 41

think of the source of the conjectural Ister as considerably farther north than the Pyrenees actually are; for the distance of the mountains from Tartessus would of course have been given in days' sailings, and we know that it was a long time before the deep indentation of the coast at the Bay of Biscay was recognized by the Greeks.[106] The western coast of Europe was thought to run generally northward, perhaps with an outward bend, and the promontory of Brittany was thought to be farther west than Cape Finisterre.

How, then, are we to imagine the map used by Herodotus? He says, as we have seen, that the Ister rises among the Celts at the city of Pyrene and that the Celts lie outside the Pillars of Herakles, neighboring the Cynetes or Cynesians, who are the last people toward the sunset of all the inhabitants of Europe. The fact that the Celts are located "outside the Pillars" again shows that they were known from the Atlantic side and presumably through the Tartessians. They are almost certainly thought of as dwelling north of the Pyrenees. These data obviously suggest a location west of the meridian of the Pillars, which would be inconceivable if the depth of the Bay of Biscay had been known or even approximately estimated. But even the Celts were not the westernmost people of Europe; for beyond them dwelt the Cynesians. We get the impression of a coast line trending northwestward from the Pillars to the Pyrenees, and thence northward with strips of land running north and south parallel to the coast beyond the Pyrenees. Just where the Oestrymnii should be located, if Berger's suggestion is accepted, cannot be determined, unless they are regarded as a Celtic tribe inhabiting a strip

---

[106] See E. H. Bunbury, *A History of Ancient Geography*, London, 1879, Vol. 1, p. 593. If Müllenhoff's reconstruction (*op. cit.*, Vol. 1, pp. 97ff.) of the old Phoenician *periplus* worked over by Avienus is accepted in this regard, it proves only that the Phoenicians early made the discovery of the true direction of the coast line; but that also is questionable.

42 THE FRAME OF THE ANCIENT GREEK MAPS

adjoining the Cynesians. If the view here suggested is true, the upper course of the Ister should be drawn from west to east and not from southwest to northeast as is commonly done in modern reconstructions of the early Greek maps. Certainly, the account of Herodotus gives no warrant for the latter direction, which would be necessary only if the Pyrenees were correctly located.

We are thus led to an approximate location for the entire northern side of the parallelogram of Ephorus and the summer tropic of the early Ionian map along a line that does not greatly depart from a direction due east and west. There remains, however, the difficulty that, according to the text of Aristotle's *Meteorology,* the Pyrenees are " toward the equinoctial sunset." There can be no doubt that there is an error here, though it is perhaps impossible to say to whom it is due. Other difficulties in this passage from the *Meteorology,* however, raise the question whether we are to think of the Stagirite as singularly careless in geographical matters or to attribute the slips to the heedlessness or ignorance of a student, whose lecture notes we may have rather than a treatise actually written by Aristotle.[107] Whatever explanation is accepted, it is certain that

---

[107] Thus we are told (*Meteorology,* 350 a 18-30) that the Asiatic Parnassus, from which the Bactrus, the Choaspes, and the Araxes rivers, as well as the Indus, take their rise, is the highest mountain "toward the winter dawn." It is inconceivable that Aristotle could have meant this, for it is impossible from any point of view. Even those who take the phrase "toward the winter dawn" as denoting direction, rather than location on a map, must find it inaccessible, as have the Greek commentators. Suggestions for the correction of the text will presently be considered, but there remains the difficulty that the mountain is called the Parnassus. As the source of the rivers in question it is clear that the reference must be to the range to which the Ionians gave the name of Caucasus and which later writers called Paropanisus. If we are in fact dealing with a student's lecture notes we may suppose that Aristotle used the latter name, which, being still unfamiliar, was mistaken for " Parnassus." This affords a possible explanation of a name that puzzled the Greek commentators. See Wilh. Stüve, *Olympiodori in Aristotelis Meteora,* in the *Commentaria in Aristotelem graeca,* Vol. 12, Part 2, Berlin, 1900, p. 104, lines 1-16. Now we shall presently see that at least from the time of Dicaearchus onward this range—the Hindu Kush—was regarded as the

(*continued on next page*)

THE NORTHERN LIMIT OF THE FRAME 43

the reading of our text, in this passage at least, is quite ancient, as is proved by the Greek commentaries [108] and especially by a passage in the *Hexaëmeron* of St. Basil, which Müllenhoff has convincingly discussed.[109] Up to a certain point this passage follows Aristotle almost verbatim, though there are additions that either come from another source or are Basil's own contribution. In the part of the passage that closely parallels the statement of Aristotle occurs this sentence: " From the *summer* sunset, at the foot of the Pyrenees Mountains, rise the Tartessus and the Ister [110]; the former empties in the sea outside the Pillars, while the Ister flowing through Europe debouches in the Euxine." This naturally raises the question whether the " equinoctial " sunset in the text of Aristotle is not a late corruption. Certainly the text is somehow at fault, because on the older maps the Pyrenees were located on the summer tropic, while, as we shall presently see, the " equinoctial sunset," or equator, was placed at the Pillars of Herakles. Some years ago, therefore, being in doubt

Continuation of footnote 107.
eastward continuation, not of the Caucasus, but of the Taurus range, and as such lay along the median axis of the Greek map. Farther south than this it was never placed; but we must infer from Aristotle's statement " when one has crossed this mountain the outer ocean is in sight " that he supposed it to lie in either the extreme north or the extreme east. In other words, he must have included in it the mountains that form the northern boundary of India. Now Eratosthenes, who followed Dicaearchus in locating these ranges on the *diaphragma*, states (Strabo, II, 1, 2) that the earlier map placed India too far to the north. This suggests that Aristotle meant either " summer " or " equinoctial " rather than " winter " dawn.

[108] Alexander of Aphrodisias (*In meteorologicorum libros commentaria*, edit. by M. Hayduck, in the *Commentaria in Aristotelem graeca*, Vol. 3, Part 2, Berlin, 1899, p. 57, lines 1, 16f.) quotes Aristotle as locating the Caucasus πρὸς τῇ θερινῇ ἀνατολῇ (toward the summer sunrise), with the variant reading τροπῇ (tropic) for ἀνατολῇ (sunrise). This shows at least how the location was understood.

[109] *Op. cit.*, Vol. 1, pp. 224ff. Basil says that the Indus in the east " flows from the winter tropic," while the Bactrus, the Choaspes, and the Araxes take their rise from midway between the sunrises. This looks like an attempted correction of the text of Aristotle as regards the Hindu Kush, which is located, after Dicaearchus and Eratosthenes, on the " equinoctial " line or equator of the Ionian map, while the headwaters of the Indus are, as in Aristotle, found in the south instead of the north.

[110] ἀπὸ δὲ δυσμῶν τῶν θερινῶν ὑπὸ τὸ Πυρηναῖον ὄρος Ταρτησσός τε καὶ Ἴστρος.

whether we should substitute " summer " for " equinoctial " or seek another remedy, I proposed [111] that the phrase " toward the equinoctial sunset " be transferred, so as to refer to the Pillars. In view of the possibility, however, that the Tartessus was the Durius (Douro) or the Tagus, which had not then occurred to me, it remains doubtful how the text is to be emended.

[111] Communicated to my friend F. H. Fobes, who mentioned the suggestion in his edition of the *Meteorology*, Cambridge, Mass., 1919, on 350 b 1.

CHAPTER IV

THE WESTERN AND EASTERN LIMITS OF THE FRAME AND THE IONIAN EQUATOR

THE WESTERN LIMIT

We must now take up the western line of the parallelogram of Ephorus. As we have observed (pp. 17f., above), it runs from the summer to the winter sunset and is entirely given over to the Celts. It will be acknowledged that this is surprising and even hardly credible; for, to mention only one circumstance, the Celts cannot have been placed south of the Pillars, whereas the " winter sunset " was far south of that point. Some change in the position of the Celts in the interval between Hecataeus (or the Phocaeans, who had close relations with Tartessus, whence the information came to him) and Ephorus is indeed readily understood, because the southward movement of the Celts fell largely in that period; but that will not account for the parallelogram, which Cosmas[112] obviously found in the text of Ephorus and which is implied in the verbatim quotation from his history. The problem seems in fact insoluble, because the general statement contained in this fragment cannot be reconciled with others that give a more reasonable account.[113] Thus Strabo[114] quotes Ephorus as saying that, according to the Tartessians, the Ethiopians extended westward to the sunset (which must, according to Ephorus' parallelogram, mean the winter sunset) and that, while some had remained there, others had occupied

[112] See above, note 38.
[113] See Müllenhoff, *Deutsche Altertumskunde,* Berlin, 1887-1900, Vol. 1, p. 82.
[114] I, 2, 26.

46 THE FRAME OF THE ANCIENT GREEK MAPS

a good part of the coast—no doubt meaning the Atlantic coast of Mauritania. Of Celts in that region there is naturally no mention. On the other hand, Strabo says that Ephorus introduced a novelty by representing Celts as living in Iberia [115]; having in mind, no doubt, that the older geographers spoke of Celts only north of the Pyrenees.[116] The text of Pseudo-Scymnus [117] also proves to be of little avail in this matter, because it is not consistent. At first [118] the Celts are said to extend from the Ethiopians (who lie along the entire southern side of the parallelogram of Ephorus) to the summer sunset; this would seem at least to place them along the entire western line as Ephorus had done, but a few verses farther on [119] we are told that they are situated opposite the Indians " at the equinoctial [120] and summer sunset, as it is said." The remaining account of the Celts is confused and admits of no consistent explanation.

It is clear that Ephorus was not consistent with himself in assigning the entire tract between the winter and the summer sunsets to the Celts. We are, however, concerned rather with the question how he came to construct that line at all and to give it, whether as a whole or only in part, to a single people. The answer must be conjectural. Perhaps the purpose was merely to complete the picture, the Celts being added to balance the Indians in the east. The older maps, of course, showed Celts in the west, but presumably only to the north of the *oikumene* proper, where

---

[115] IV, 4, 6. Celts are said to extend as far as Gades.
[116] Strabo (IV, 1, 14) says the name of Celts was first given to the people of Narbo and later extended by the Greeks to include all the Gauls.
[117] A versified *periplus* dating probably from the first century before Christ, formerly attributed to Scymnus, who wrote a century earlier. The text may be found in Carl Müller, *Geographi graeci minores,* Paris, 1882, Vol. 1, pp. 196ff.
[118] Pseudo-Scymnus, v. 173f.
[119] *Ibid.,* v. 176ff.
[120] Müller suggests ὑπὸ χειμαρινῆς. If this reading is adopted the limits assumed are those of the parallelogram of Ephorus.

they may have been assigned to a meridional strip near the ocean. By the time Ephorus wrote there had begun in the west a southward thrust of northern peoples, chief among whom were the tribes regarded as Celts. How far these may have penetrated Spain by that time we do not know, but it is entirely possible that a Greek might reasonably have thought that the Celts had won the whole of Spain. Furthermore, while the bordering peoples were from an early date regarded as living along the limits of the *oikumene,* it was known that they had here and there thrust themselves inside the frame. The Cimmerians had invaded Asia Minor, the Scythians had penetrated south of the Ister and had made inroads into Media. Particularly at the extremities or corners of the frame was there a certain overlapping. Scythians were known in northern India, and the " Ethiopians " of southern India were as familiar to Aeschylus as to Herodotus. In the west, also, as we have seen, Ephorus recognized a northward movement of Ethiopians from their proper sphere into Mauritania along the Atlantic seaboard, and the " brow " of Herodotus, with its bordering peoples, was supposed to extend across Libya from Egyptian Thebes toward the Pillars. Thus the parallelogram, in any case, appears to have been merely a diagrammatic scheme, which, however, preserved substantially the frame of the Ionian maps.

The Eastern Limit

It is hardly necessary to dwell at length on the eastern side of the parallelogram, where the Indians were set down as extending from the summer to the winter sunrise. As we have observed, India was always regarded as bounded on the north by the chain of lofty mountains of which the Himalayas are the highest. From the time of Dicaearchus and Eratosthenes these mountains were conceived to be an eastward extension of the Taurus range, lying along a

48 THE FRAME OF THE ANCIENT GREEK MAPS

line due east of it; but Eratosthenes remarked that the older maps placed them much farther north. Every indication points to the conclusion that the early geographers, chief among them Hecataeus, regarded these mountains as in line with the Caucasus, which was first known to the Greeks northeast of the Euxine. Whether Anaximander on his map attempted to sketch the *oikumene* very far eastward of the Phasis we do not know; we may be sure, however, that the " marvellous improvement " [121] on his predecessor's map made by Hecataeus was chiefly in the east, for which region Hecataeus long continued to be recognized as the principal authority. Wherein his improvement consisted in detail we cannot say, but it is certain that he had the benefit of the expedition of Darius to India and of the account of the voyage of Scylax. From the latter he could have gathered some conception of the southward extension of India, which he doubtless must have placed roughly on the parallel of southern Arabia deemed by Herodotus to be the southernmost land of the *oikumene*.[122] This limit again coincides with the winter tropic of Herodotus.[123] Thus we obtain, with reasonable certainty, for the map of Hecataeus the same limits as those adopted by Ephorus in his parallelogram.

The location of the eastern lands with reference to the Ionian tropics long continued, though one significant change was made. Deïmachus, who lived early in the third century, stated that India lay between " the autumnal equinoctial and the winter sunrise," thus transferring the northern boundary of India farther south to a position better in accord with information derived through the campaigns of

[121] Agathemerus, I, 1.
[122] III, 107.
[123] The westward extension of this line is clearly suggested by Herodotus (III, 114): ἀποκλινομένης δὲ μεσαμβρίης παρήκει πρὸς δύνοντα ἥλιον ἡ Αἰθιοπίη χώρη ἐσχάτη τῶν οἰκεομένων—that is to say, Ethiopia extended westward from the true (absolute) south, which obviously lay roughly on the parallel of southern Arabia. See above, note 53.

THE WESTERN AND EASTERN LIMITS 49

Alexander. Strabo,[124] who reports the statement, agreed with Eratosthenes in calling Deïmachus an ignoramus in astronomy for saying " autumnal," since there is no difference between the autumnal and the vernal equinoctials, but apparently Strabo did not otherwise object to Deïmachus' delimitation. Now Deïmachus was a contemporary of Dicaearchus, who located India between his median axis, or *diaphragma,* and the parallel of Meroë. The western end of the *diaphragma* of Dicaearchus ran through the Strait of Gibraltar on the equator of the Ionian map. Whether or not Dicaearchus was the first to draw this line to the northern boundary of India, we suspect that Deïmachus' statement was derived from Dicaearchus. Being undoubtedly, as Eratosthenes and Strabo say, an ignoramus in matters of astronomy, it is not likely that Deïmachus invented the formula. If, then, as it seems, he was in fact reproducing Dicaearchus, only with the stupid addition of " autumnal," we may assume that Dicaearchus, a pupil of Aristotle, used somewhat similar terms to those of Ephorus. I am inclined to take this view, which, if accepted, has an important historical bearing; for it must follow that the map of Dicaearchus did not differ essentially from that of the Ionians. That Dicaearchus did in fact follow in essentials the Ionian map is perhaps to be inferred from the circumstance that Polybius, who claimed to disregard the older geographers and to address his criticisms to Dicaearchus and Eratosthenes, actually used the formulas of the Ionians, which were of course discarded by Eratosthenes. Thus Polybius was clearly using the old Ionian terms when he stated [125] that Libya lay between the Nile and the Pillars of Herakles, lying to the south and continuing to the winter sunset, thence up to the equinoctial sunset, at the Pillars. Such a description,

[124] II, 1, 19 (after Eratosthenes).
[125] Polybius, III, 37.

50 THE FRAME OF THE ANCIENT GREEK MAPS

although an anachronism for a geographer living after Eratosthenes, was natural to one who used the older maps.

There is no reason to inquire further regarding the eastern line of the parallelogram of Ephorus, because it was obviously intended to be no more than a part of a diagram. The Indians, like the Celts, Ethiopians, and Scythians were one of the nations who dwelt outside of the parallelogram. The western border of India proper, hence the eastern limit of the parallelogram, was the Indus. Later maps drew the course of this river from north to south, but we know that it was not so represented on the old maps, which were certainly used by Ephorus; for Strabo [126] says they drew it from northwest to southeast. In the east, therefore, as in the west, the meridional coördinate of Ephorus is quite arbitrary.

ORIGINS OF HERODOTUS' GEOGRAPHICAL SKETCH OF ASIA

We are sure that the map of Hecataeus embodied the results of the expedition of Darius to India, and it is altogether probable that it likewise utilized what Professor Myres has aptly called the "Persian map." The latter may be reconstructed from the geographical sketch of Asia given by Herodotus.[127] We may assume that the attempt

[126] II, 1, 34. Hipparchus here, as at many other points, followed the old maps as against Eratosthenes, who drew the course of the Indus from north to south.

[127] Herodotus, IV, 37-41. There is, so far as I know, no direct evidence of actual maps made in Persia, but I do not doubt that they existed. The fact that a sketch map of the Euphrates valley of Babylonian origin has been found (see above, note 6) makes this probable, and Herodotus (III, 136) tells of a "description" of the Greek coasts made for Darius. That this description included at least a sketch map cannot be doubted; and it would not be necessary to explain the latter by the fact that the expedition was headed by the Greek Democedes. The expedition set out from Sidon and was certainly manned, at least in good part, by Phoenicians, who unquestionably were familiar with sketch maps of the Mediterranean, though they knew the African coast better than the European. It seems to me all but certain that it was Hecataeus who combined the Persian with the Ionian map. On Herrmann's suggestion regarding a Phoenician map, dated about 1000 B. C., see above, note 72.

THE WESTERN AND EASTERN LIMITS 51

to incorporate the results of the voyage of Scylax produced an eastward extension of the map. Apparently the principal lines adopted by Hecataeus followed those already established by Anaximander, the northern running east from the Caucasus and the southern approximating the parallel of the southern borders of Arabia. Just what results followed from Hecataeus' inclusion of the " Persian map " we cannot say; but since the Caucasus was the northern boundary of the Persian Empire, which extended to the southern sea, it seems likely that the main Ionian parallels were not affected, unless Hecataeus located his Caucasus with reference to the southern end of the Caspian, which it skirts, thus placing the northern parallel not quite so far north as had previously been the case. The geographical sketch in Herodotus, indeed, suggests this; for in western Asia it traces two important lines, the first beginning at the Phasis and extending to the Hellespont, the second beginning at the Myriandric Gulf (Gulf of Alexandretta) and running along the coast of Cilicia to the Triopian headland (Cape Crio). Between these lines lies Asia Minor, which Herodotus conceived to be much narrower than it actually is.[128]

It will repay us to consider this geographical sketch somewhat more in detail. Herodotus indicates a meridional belt extending northward from the Indian Ocean and including, after the Persians, the Medes and then the Saspires and the Colchians, who dwell by the Euxine. Westward of this meridional belt he describes two longitudinal tracts, one of which includes all Asia Minor, its southern boundary running from the Triopian Cape to the Myriandric Gulf. What portion of the meridional belt this longitudinal tract was supposed to face is not expressly stated but may be inferred from the fact that the second of the longitudinal

[128] According to Herodotus (II, 34) the distance from Cilicia to Sinope is supposed to be such that a man traveling light might traverse it in five days.

52 THE FRAME OF THE ANCIENT GREEK MAPS

tracts is made to begin with Persia. We may assume, therefore, that the line from the Triopian Cape to the Myriandric Gulf, when produced eastward, was supposed to follow the boundary between Persia and Media. Actually such a line would be reasonably accurate. Between the Myriandric Gulf and the boundary separating Persia from Media such a line would coincide with the " Taurus " and the *diaphragma* of the charts of Dicaearchus and Eratosthenes. There is every reason, therefore, to think that the Ionian map of the fifth century showed the " equator " or median longitudinal axis from the Pillars to the eastern boundary of Persia.

It is not without interest to compare with this sketch in Herodotus the list of subject peoples given in the inscriptions on the tomb of Darius at Behistun.[129] Here, besides a line running eastward from Persia and Media and extending from Elam to India and the Sacae, two series of regions are mentioned, the first including Armenia, Cappadocia, Sardis, and Ionia, and the second Babylon, Assyria, Arabia and Egypt. This arrangement agrees remarkably with that of the tracts (ἀκταί) described by Herodotus and almost certainly by Hecataeus before him. It does not necessarily follow that the original, which was undoubtedly Persian, took the form of a graphic sketch, but we may naturally suppose that it did; and the schematic or diagrammatic form indicated in the texts agrees very well with that to be inferred from the journey from Ionia to Susa that Aristagoras sketched for King Cleomenes of Sparta, indicating the route on an Ionian map that he had brought with him.[130] Aristagoras had consulted Hecataeus

[129] See F. H. Weissbach, " Die Keilinschriften am Grabe des Darius Hystaspis," in *Abhandl. der Sächs. Gesell. der Wiss.*, Leipzig, *Philolog.-Hist. Klasse*, Vol. 29, No. 1, 1911. W. H. Roscher ("Das Alter der Weltkarte in ' Hippokrates ' περὶ ἑβδομάδων und die Reichskarte des Darius Hystaspis," in *Philologus*, Vol. 70, 1911, pp. 529-538) referred to the inscription as presupposing a map, but overlooked the corresponding account of Herodotus.

[130] Herodotus, V, 49.

THE WESTERN AND EASTERN LIMITS

regarding the advisability of revolting from Persia,[131] and it may well have been the map of Hecataeus that he showed to Cleomenes.

THE IONIAN EQUATOR

The outer frame being thus established, we must now inquire whether there were other lines of fundamental importance and of general acceptance. In our sources are given many indications of lines drawn for special purposes and useful in locating particular places or peoples, but there is perhaps only one, the equator of the Ionians, that had much significance in relation to the map as a whole. We have pointed out that the Ionians conceived the earth after the analogy of the horizon, on which the extreme points of the summer and winter risings and settings of the sun marked the limits of the *oikumene*. These limits were regarded, however vaguely, as those of habitability by reason of heat and cold; in other words, they were related to climate, and consequently the tract between them might in a sense be considered a zone. It need hardly be emphasized that such a zone, if one chooses to call it so, was something quite different from our temperate zone, because it was determined on different principles. The facts of experience, however, are not altered by theory, and much the same things, within limits, could be said of the habitable tract of the Ionian map as of the temperate zone of our maps. The climate of Asia Minor was thus praised as the ideal; but the reason given for its character was, as we have seen, that it lay midway between the sunrises—in other words, along the equator of the Ionian map.

This line lies along the longitudinal axis of the Mediterranean, which was the core of the Ionian map. Professor Myres calls attention [132] to numerous passages in

---

[131] Herodotus, V, 36.
[132] J. L. Myres, "An Attempt to Reconstruct the Maps Used by Herodotus," in *Geogr. Journ.*, Vol. 8, 1896, pp. 607-609.

Herodotus that indicate the location of places symmetrically north and south of this line. This fact proves that it was the main axis of the maps; and, whether it was adopted by the geographers from earlier portulan charts or not, it was certainly old. Presumably the earliest maps did not extend far beyond the eastern end of the Mediterranean. On them the equator would accordingly run from the Pillars of Herakles along the axis of the Mediterranean to Asia Minor and thence along the Taurus eastward. When the "Persian map" was incorporated, as we have seen, the southern limit of the longitudinal tract, indicated by Herodotus as comprising Asia Minor, followed the same line eastward, to the eastern border of Persia. After the expedition of Alexander to India the Hindu Kush and the Himalayas, formerly regarded as an extension of the Caucasus, were located on the parallel of the Taurus. The line thus revised and produced was adopted by Dicaearchus and Eratosthenes as the *diaphragma,* or median axis of their maps. The Himalayas, which lay along this axis, were then, as now, considered the northern boundary of India. It is, therefore, significant that Deïmachus, a contemporary of Dicaearchus, placed the northern limit of India on the (autumnal) " equinoctial sunrise," while Polybius,[133] who still occasionally used the maps and the language of the Ionians, placed the Pillars of Herakles at the " equinoctial sunset."

There can be no doubt, therefore, that the Ionian map had an " equator " based, like its " tropics," on the appearance of the fixed horizon with the points marked by the rising and settings of the sun. Indeed, the " equator," it would seem, must have been especially important as the main line of reference in the world known to the Greeks; for otherwise it is hard to understand why Eratosthenes should have retained the line as the main longitudinal axis

---

[133] III, 37; see above, pp. 49f.

THE WESTERN AND EASTERN LIMITS

of his own map, constructed on other principles. Obviously he was following an established tradition.

Now that we can picture to ourselves the outlines of the Ionian map, we see that it was essentially diagrammatic. It was probably in the earlier examples circular, as Herodotus and Aristotle say; but much of the circle, which was suggested by the horizon, was given over to parts unknown, lying beyond the limits of the *oikumene*. One might play with that portion as one's fancy suggested; but the inner map, with which the geographer was chiefly, indeed almost exclusively, concerned, was doubtless severely regular, with a marked tendency to subdivision by straight lines. Professor Myres fully explains this tendency, and there are abundant indications of these lines in the geographic literature. Since it is not our purpose to attempt a reconstruction of the ancient maps in detail, we need not go into this subject further. It must suffice to say that these lines, if drawn on the true map, would not be straight but would everywhere show distortion. Consequently they would not at all resemble the parallels and meridians on a modern map, though they were intended to serve the same purpose—that of aiding the location of particular places or peoples in relation to others.

CHAPTER V

OBSERVATIONS ON WHICH THE FRAME WAS BASED

We naturally ask how the Ionians came to construct their maps in this way. Before general maps were attempted there were no doubt sketches of routes between particular points, and such sketches could readily be extended and combined. These would necessarily take account of directions from point to point [134]; but since the successive points were presumably at first not far apart, the instructions would not have to be very exact. The directions north, south, east, and west were known, but intermediate points were not indicated (*e. g.*, in Homer). With these so-called cardinal points there was given the possibility of a system of coördinates, but not an actual frame. The horizon, within which the frame of the Greek map must fall, was believed to be fixed and not changing with the station of the observer; hence, if there was to be a frame, it had to be related to fixed points on the permanent horizon. Directions to be indicated were accordingly regarded as absolute.[135] Homer knew the winds Boreas, Eurus, Notus, and Zephyrus; later other winds were added, each with its specific direction, until a complete wind rose was constructed that would have served almost

---

[134] E. g., *Odyssey*, III, 169-172.
[135] A. Rehm ("Antike Windrosen," in *Sitzungsber. Münchener Akad. Philolog.-Hist. Klasse,* 1916, No. 3) innocently says that the locations of certain places by Aristotle and Polybius, which he supposes were made with reference to the wind rose, are quite correct, though one misses an indication of the point from which the direction was taken! The reason why no point was indicated is that the reference was not to mere direction, which would of course change with the observer's position, but to a location regarded as absolute. See above, note 41.

as well as a compass card. There is, however, no evidence that it was actually used for geographical purposes, as is often assumed. The evidence points to the conclusion that the wind rose, like the map, was suggested by the horizon and was at first divided by lines drawn toward the risings and settings of the sun; for the north and south Boreas and Notus served, until true north and midday took their places.

As the frame of the map was based mainly on observations of the sun, we might naturally conclude that the Ionians used the gnomon or sundial as their model for the map, especially since tradition attributed the invention of the gnomon to Anaximander, to whom it likewise ascribed the first general map of the earth. It must be admitted that the suggestion is plausible; and we cannot read the modern histories of geography without seeing that in countless connections it is tacitly assumed that the ancients used the gnomon in some way for geographical purposes. Unfortunately, there is, so far as we know, no trace of such use until after the acceptance of the sphericity of the earth in the fourth century, when the gnomon was employed in measuring the midday shadow of the stylus at the solstices.[136] Surprising as this may at first appear, reflection will perhaps make it intelligible. The gnomon was undoubtedly employed from an early date, but, so far as we know, at first only as a timepiece. It readily marked not only the daylight hours but also the longest and shortest days, or the solstices, as well as the equinoxes, and to this extent may be said to have had astronomical uses. We may, however, question the ac-

[136] The passage in Herodotus (IV, 42) in which is described the circumnavigation of Africa by Phoenicians under the orders of Necho, King of Egypt, could hardly be cited as an earlier instance of taking the position of the sun in determining latitude. It is true that the Phoenicians are said to have reported that, as they rounded the southern coast of Africa from east to west, the sun was on their right hand; but Herodotus did not believe the report and obviously saw no significance in the fact, if it were accepted.

curacy of the statement that Anaximander, who seems to have introduced the gnomon, discovered (with or without it) the inclination of the ecliptic [137]; this implies more than we can well allow. All the Ionians accepted the doctrine of the dip of the earth to the south; but that is not quite the same thing. Such use as the Ionians made of the gnomon would not affect their theory of the shape of the earth; for, since the apparent motion of the sun depends solely on the varying angle of its course, the lines traced by its shadow tell nothing about the shape of the earth. When the sphericity of the earth was accepted the network of the dial did of course assume added significance; and, since that conception was essentially a corollary of the spherical vault of the heavens, we can readily understand why Aristarchus of Samos (*ca.* 260 B. C.) devised the *skaphe*,[138] a dial constructed in a hemispherical bowl, showing in reverse the passage of the sun across the sky.

All this, however, had no obvious bearing on geography. Yet, in view of the undoubted relation of the Ionian map to the horizon and the points that were marked on the dial of the gnomon, it is not difficult to understand why historians of Greek geography should be constantly thinking of that instrument. The fact remains, however, that there is no evidence that it was employed for geographical purposes before the latter half of the fourth century; and we may affirm with some confidence that the absence of evidence is not due to chance, for the observations that can be made with this instrument can be made equally well without it. The portable gnomon was, therefore, at first probably regarded merely as a novelty, especially since it must have seemed both inconvenient and inaccurate. Primitive peoples the world over have observed the changes in the risings and settings of the sun and have marked them

---

[137] Pliny, *Naturalis historia* (II, 31). The statement apparently goes back to Posidonius, of whose unhistorical point of view we shall presently speak.
[138] Vitruvius, *De architectura*, IX, 8, 1.

BASIC OBSERVATIONS

by prominent objects on the horizon.[139] By close observation they can even tell in this way the number of days that will elapse before the solstice or the equinox: with a small portable dial this would be impossible. A passage in the *Odyssey*,[140] where the island Syrie is mentioned as the place where the sun turns, probably refers to this use of points on the horizon and suggests how the solstices and the equinox, observed on the horizon, were given absolute geographical positions, which, when referred to the *terra cognita*, were located on the limits of the *oikumene*. The extension of the frame from a restricted horizon to the *oikumene* was the more easily made because the heavens were supposed to owe their apparently spherical form to distance and because within the limits known to the Greeks the shadow of the gnomon would not greatly change, except as the sun stood higher or lower in the heavens as one went north or south—a fact sufficiently explained for them by the notion that the sun, conceived as not very distant, really stands over central Africa in winter and approaches Scythia in summer.[141]

[139] See Martin Nilsson, *Primitive Time-Reckoning*, Lund, 1920, for the evidence. Hesiod (*Works and Days*, 564, 663) dates the time for pruning vines by reference to the winter solstice and the rising of Arcturus, and the season for sailing by reference to the summer solstice. The solstices could readily be marked by points on the horizon.

[140] XV, 403f. The passage was correctly interpreted by T. H. Martin in Charles Daremberg and Édouard Saglio, *Dictionnaire des antiquités grecques et romaines*, Paris, 1873-1919, Vol. 1, p. 477a.

[141] Herodotus, IV, 36; Hippocrates, *De victu*, B 37-38 (Emile Littré's edit., Paris, 1839-1861, Vol. 6, pp. 528-530); *De aëre*, 19 (Littré's edit., Vol. 2, p. 70).

PART II
THE SPHERICAL EARTH IN RELATION TO THE FRAME

CHAPTER VI

THE SUPPOSED EARLY DISCOVERY OF THE SPHERICITY OF THE EARTH

We should like to know when and by whom it was first suggested that the earth, which at first was regarded as a flat disk, was really a sphere; nor was it left for modern scholars to raise this interesting question. The Greeks themselves, at least as early as the sixth century before Christ, raised such questions as this, having become aware that things had not always been as they were. Besides attributing the origin of many things to various gods they conceived of a number of " culture heroes," like Prometheus and Palamedes, from whom they derived important " inventions." As the historical sense developed, search was made among such records as existed for the first evidence of institutions, arts, practices, devices, and ideas. When a philological interest awoke, the question came to be asked also who first introduced the use of a word in a given sense. Thus we find in our sources many suggestions regarding the discoveries that seemed important to later ages. We know by abundant evidence in our own times how difficult it is to determine such matters, because beginnings are apt to be inconspicuous and because ideas attract attention only when they take tangible shape or assume practical importance. Statements about the origins of ideas are likely to be mere guesses that we should evaluate according to the competence of those who offer them.

Character of the Source Materials

In regard to the discovery of the sphericity of the earth we have a number of statements, the value of which can be rightly appraised only by considering the character of our sources. The older modern literature dealing with this and other kindred subjects is rendered practically useless by the uncritical acceptance of statements, irrespective of their sources, provided that they fell in with the preconceived notions of the historians. It was only in the last quarter of the nineteenth century that a fruitful study of all the available documents was begun, and it need hardly be said that much still remains to be done in this direction and that even now absolute certainty cannot be claimed for our conclusions.

In Part I of this study we dealt for the most part with writers who were themselves using, or at least familiar with, the early maps drawn by those who conceived the earth as a flat disk. The principal questions arising in that connection were whether those writers accurately represented the maps. In our present inquiry we are most concerned to know how far we may trust our sources. The question is unfortunately complicated by a number of facts, the most important of which is that, with the exception of Plato and Aristotle, the authors whom we have to cite are preserved only in fragments until we come to quite late times, when in Diogenes Laërtius we have a hotchpotch of statements gathered from all quarters and further confused by careless editing. To reduce this *indigesta moles* to some semblance of order is a task of great difficulty, not yet fully achieved. The literature excerpted and combined by Diogenes Laërtius is of several kinds, one of which is biographical while another is concerned with scientific and philosophical doctrines.

We need not here discuss all the lines of tradition that meet in these late handbooks, but we must speak briefly of that which deals with the history of scientific and philosophical doctrines. Where and when this tradition started it is perhaps impossible to determine. We see that certain Sophists of the fifth century were interested in the history of thought, and there is clear evidence that the older historico-geographical literature of the fifth (possibly sixth) century dealt at least occasionally with scientific ideas and their exponents. Xenophon, Plato, and Isocrates in the first half of the fourth century concerned themselves with these matters in a way showing that a tradition had already been forming. In the latter half of the fourth century Aristotle in frequent introductory statements summed up previous opinions on the questions he was discussing. Then his pupil and successor as head of his school, Theophrastus, wrote a comprehensive account of the *Opinions of the Physical Philosophers,* which, except for a part that does not at present concern us, survives directly in meager fragments. This work was a godsend, as Burnet says,[142] to later epitomators and compilers of handbooks, who handed on extracts from it by various subordinate lines of tradition. Professor Hermann Diels in his important *Doxographi graeci,* Berlin, 1879, following earlier investigations by Usener and others, made a profound study of this "doxographic tradition" and showed the relation of the several lines to one another and to the original work of Theophrastus.[143]

It is not necessary to set forth in detail the results of his study; we must, however, point out a fact fully established by Diels: the most influential epitome, made by Aëtius about 100 A. D., derives directly from a similar work, dated a century and a half earlier, a work that Diels called *Vetusta placita* but later defined more accurately as the *Posidonian areskonta* because it is plainly influenced strongly by Posidonius, a teacher of Cicero. This fact, to

[142] John Burnet, *Early Greek Philosophy,* 3rd edit., London, 1920, p. 34.
[143] A diagram showing the relation of the sources may be conveniently consulted in Sir Thomas Heath, *Aristarchus of Samos* . . . , Oxford, 1913, p. 3.

which too little attention has been paid, is for several reasons of prime importance in connection with the question of the sphericity of the earth. First of all, several of the statements that refer the discovery to quite early thinkers come from Aëtius and, therefore, indirectly from the circle of Posidonius, while others come from those parts of the text of Diogenes Laërtius that have a very dubious origin, possibly the source used by Aëtius. A further reason for scrutinizing these reports lies in the character and associations of Posidonius.

Misleading Ideas Probably Due to Posidonius

Posidonius of Apamea (born about 135 B. C.) was a man of notable attainments in many directions, and he exercised a vast influence, the limits of which are difficult if not impossible to determine. He was, however, notable for his warm personality, his learning, and his eloquence, rather than for judgment and historical criticism. His want of critical judgment, indeed, was not peculiar to himself. He was a Stoic, and the Stoics, beginning with Zeno, the founder of the school, were especially prone to read their own ideas into the works of their predecessors. It is well known that they regarded Homer as the wise man in whose poems one might discover a knowledge of almost all that was known in their day. Eratosthenes was at great pains to refute the claims of the Stoics that Homer knew almost everything about geography, and Strabo in characteristic Stoic fashion criticised Eratosthenes for so doing. In this, Strabo was almost certainly following Posidonius. One scholar, indeed, with Stoic leanings, Crates of Mallos, chief librarian of Pergamum and rival of the great Aristarchus of Alexandria, discovered in Homer evidence of the sphericity of the earth, as Zeno, the founder of the Stoa, discovered it in Hesiod. The Stoics practiced

THE SPHERICITY OF THE EARTH

what Cicero [144] called "accommodation"—that is to say, the assimilation of earlier doctrines to their own. That practice was destined to lead to great carelessness in historical matters, especially when combined with the tendency to which the Stoics were prone, of discovering in plain language hidden meanings to be brought out by allegorical interpretation. We shall later find that Posidonius asserted that Parmenides knew the geographical zones; he must therefore have credited Parmenides with believing the earth to be spherical. Since no good reason exists for thinking that the philosopher held either of these views and the best of reasons can be given for holding that Posidonius was in error, our suspicion is confirmed that Posidonius was at least the proximate source for the statements of Aëtius that Thales and the Stoics and their respective adherents taught the sphericity of the earth and for the assertion of Diogenes Laërtius that Anaximander also held that doctrine.[145]

The Sphericity of the Earth Unknown to the Early Philosophers

No one with the least critical judgment now believes that either Homer and Hesiod or Thales and Anaximander had the remotest suspicion of the rotundity of the earth. Indeed, we learn from better sources that Anaximander and his successor, Anaximenes, regarded the earth as approximately flat,[146] as we should expect in view of the whole course of early Greek thought. It happens that there is

---

[144] *De natura deorum*, I, 15, 41.
[145] Aëtius, III, 10, 1; Diogenes Laërtius, II, 1.
[146] For Anaximenes, see Aristotle (*De caelo*, 294 b 13) and Hippolytus (*Refutatio*, I, 7, 4); for Anaximander, Hippolytus (I, 6, 3), where it is distinctly implied that the earth is shaped like the drum of a column (i. e. though circular, it has flat surfaces above and below). I believe that a corruption of the text is responsible for the comparison of the earth with the drum of a column (see below, note 173); but, even so, Anaximander must have regarded the earth as essentially flat.

good authority also for the statement that the other philosophers who may be grouped with the early Ionians—that is to say, Anaxagoras, Archelaus, and Democritus—knew nothing of the sphericity of the earth.[147] As we have before pointed out, the view that the earth is a circular disk was, no doubt, at first suggested by the horizon; but for the Ionians it was certainly confirmed by the conception, shared by all of them, of a cosmic vortex, which originally circled about the earth as its center and was in the same plane with the earth. Aristotle, indeed, says that this view was held by all who thought of a cosmos as originating.[148] This would certainly apply to all the Ionians and might even be thought to apply to Pythagoras.[149] The vortex, suggested as it was by the phenomena of whirlwinds and whirlpools, would naturally lend support to the view that the earth was a disk. So far as we know, Plato[150] was the first to recognize that a spherical earth might be similarly poised in the center of a cosmic sphere and require no support. The Ionians, however, did not think of the cosmos as spherical but as a series of circles, which, by the dipping of the earth disk at the south, had come to wheel about it at an angle that was greater in summer than in winter.

The view expressed by Plato, whether original with him or not, was obviously a new application of an older conception connected with the notion that the earth is a flat disk. The phenomena of whirlpools and whirlwinds would not naturally suggest a sphere as the form either of the cosmos as a whole or of the earth at its center. One cannot well accept the report of Aristotle[151] that Anaxi-

---

[147] Aristotle (*De caelo*, 294 b 13-14), for Anaxagoras and Democritus; Hippolytus (I, 9, 4), for Archelaus.
[148] *De caelo*, 295 a 10-14.
[149] Aristotle, *Metaphysics*, 1091 a 18.
[150] *Phaedo*, 108-109 a.
[151] *De caelo*, 295 b 11-16. It seems probable that Aristotle mistook an innocent circumstantial participial phrase, intended merely to describe the posi-
(continued on next page)

THE SPHERICITY OF THE EARTH

mander explained the fact that the earth does not fall by the circumstance that it lies equidistant from the periphery (obviously of a spherical cosmos). There is in fact nothing to suggest a spherical cosmos in the scheme of Anaximander: everything points rather to a system of concentric circles. We may be sure, therefore, that Aristotle here is guilty of the confusion, so common in our sources, between the circle and the sphere.

Nor is it probable that Anaximander had raised the question why the earth does not fall. We are told,[152] to be sure, that his predecessor, Thales, held that the earth rested on water. If there was any ground for this assertion we may be sure that " earth " referred not to the terrestrial body but to the land mass, between which our sources do not always distinguish. Anaximenes seems to have thought that " air " enveloped all things, therefore also the earth. But one may doubt, again, whether Aristotle was justified by unambiguous statements when he stated [153] that Anaximenes, Anaxagoras, and Democritus explained the fact that the earth does not fall by its broad, flat form, since a body of that sort, acting as a lid, is supported by the air beneath it, which cannot escape. Just when the problem why the earth does not fall was raised we cannot be sure, although it is certain that it was much discussed in the latter part of the fifth century before

Continuation of footnote 151.
tion of the earth as equidistant from all points of the encircling heavens, for a causal use of the participle and inferred that Anaximander meant to explain why the earth does not fall. This interpretation was presumably current in the school of Plato.

[152] Aristotle, *De caelo*, 294 a 28-33. Aristotle here refers to the experiment of Empedocles with the waterclock, which shows that air may support a column of water, but we do not know that Empedocles drew inferences from it regarding the earth. The other experiment of Empedocles with water in a container that was whirled round had to do with centrifugal force, which, if applied to the cosmos, would explain the phenomena of a vortex in a plane, but not in a spherical universe.

[153] *De caelo*, 294 b 13-21.

Christ.[154] That it was connected somehow with the conception of the cosmos as a sphere is suggested by the remarks of Plato, to which reference has been made.

THE CLAIMS OF PARMENIDES

When and by whom the sphericity of the cosmos as a scientific conception was first suggested is difficult to determine. As has been pointed out, the Ionians disregarded the apparent hemisphere of the heavens and conceived of the world as a series of concentric circles of varying breadth. Parmenides pronounced the "All" a perfect sphere, and Empedocles, about the middle of the fifth century, likewise spoke of the Sphere in which through the action of Love the elements separated by Strife were united. One infers that this was suggested by aesthetic considerations, the beauty and perfection of the sphere being its chief recommendation. Whether the conception was applied to the cosmos and its parts is more than doubtful. The description of the "crowns" by Parmenides[155] certainly

[154] Aristophanes (*Clouds*, 264) apostrophizes "Air, that dost support the earth on high." Euripides (*Trojan Women*, 884) hails Zeus (doubtless conceived as the sky) as "the support of the earth." These references date from the last quarter of the fifth century. Hippocrates (*De flatibus*, 3; Littrés edit., Vol. 6, p. 94) repeats the phrase in reference to the air. This work likewise may be dated about 400 B. C. It is not so easy to assign a date for the pseudo-Hippocratic treatise *De septimanis*, 2, which represents the earth as situated at the center of the cosmos and borne by air. Since the context refers to the antipodes, it is assumed that the author regarded the earth as spherical. Roscher, as is well known, made much of this treatise, contending that it was the oldest document of Ionian science, even earlier than Anaximander. Few if any scholars now accept his views. Franz Boll (*Das Lebensalter* . . . , Leipzig and Berlin, 1913, pp. 54-58) and Abel Rey (*La jeunesse de la science grecque*, Paris, 1933, pp. 437-438) date the pseudo-Hippocratic treatise *ca.* 450 B. C., because they believe that Parmenides held the earth to be spherical and suppose that the writer of that work depended on him. Since I see no reason to accept their view regarding Parmenides and there is much in the treatise that suggests dependence on Diogenes of Apollonia, I should date this problematic work not earlier than the last quarter of the fifth century and am inclined to think it was written even later, by one who combined the general philosophy of Diogenes with the conception of the sphericity of the earth.

[155] Fr. 12, in Hermann Diels, *Die Fragmente der Vorsokratiker*, 5th edit., Vol. 1, Berlin, 1934.

THE SPHERICITY OF THE EARTH 71

suggests that he, like the Ionians, conceived of the heavenly bodies as disposed in circles, and there is no reason to think that Empedocles, who was clearly much indebted to Parmenides, regarded the earth otherwise than as flat. The doxographic tradition attributes the discovery of the obliquity of the ecliptic variously to Thales and to Pythagoras [156]; but we know that the Ionians one and all explained the angle of the sun's course by the "dip" of the earth at the south. It would seem to be significant that the doxographic tradition says [157] that, though Pythagoras had discovered the obliquity, Oenopides of Chios, who flourished in the latter half of the fifth century, claimed the discovery as his own. In view of the character of the legends about Pythagoras we are inclined to think that the claim of Oenopides was valid; and this conclusion is amply justified by the fact that Eudemus, the pupil of Aristotle who studied and wrote the history of the mathematical sciences, is quoted [158] as saying in his *History of Astronomy* that Oenopides first discovered the obliquity of the zodiac.

What does his discovery imply? Obviously it was no longer a question solely of the changing positions of the sun, which might be explained by the "dip" postulated by the Ionians. The zodiac must have been defined by taking account not only of the sun, which seems to traverse the ecliptic, but also of the planets, which, by reason of the different angles of their orbits to the ecliptic, wander through a broad belt. Oenopides, then, we must infer, was concerned with the changing positions of the planets relative to the stars; and his other interest, shown in the attempt to define the length of the "great year," gives evidence of his occupation with the allied problem of the solar year and its multiples. Eudemus, who himself

---

[156] Aëtius, II, 12, 3.
[157] Aëtius, II, 12, 2.
[158] Theo Smyrnaeus, *On the Mathematical Knowledge Which is Needed to Read Plato,* Eduard Hiller's edit., Leipzig, 1878, p. 198, line 14.

placed the tropic at 24° from the equator,[159] must be assumed to have had adequate knowledge of the work of his predecessors; and we may conclude that it was in the last quarter of the fifth century that astronomers made observations sufficiently exact to render possible the location of the ecliptic and the limitation of the zodiac. This does not imply, however, that the heavenly bodies had not earlier been assigned to belts of varying breadth, but only that there were no exact determinations. If we may in this respect trust the doxographic tradition,[160] Parmenides, like Anaximander, broke up the vault of heaven into bands and represented them as crossing one another; the precise meaning of these celestial bands is doubtful, but if there is a reference to the zodiac it is extremely vague. It seems more reasonable to think primarily of the Milky Way. In any case, the description of the cosmos contained in Parmenides' *Opinions of Mortals* affords no justification for identifying the cosmos, consisting of a system of circles, with the "All," which Parmenides declared to be a perfect sphere. In his cosmography Parmenides, then, may be regarded as essentially following the Ionians, and the natural inference would be that he shared also their conception of the earth as a flat disk.

The Claims of Pythagoras

We have now to consider the claims of Pythagoras. It is now generally admitted that the tradition regarding Pythagoras is utterly untrustworthy, because he soon be-

---

[159] *Ibid.*, p. 199. Sir Thomas Heath (*Aristarchus of Samos* . . . , Oxford, 1913, p. 131, note 4) supposes that the determination may have been made by Pythagoras; but he presents no evidence and there is in fact no reason to think that he is right.

[160] Aëtius, II, 7, 1. The bands were of course indicated by Parmenides, but whether they were merely "one over the other" or were "crossing," as Burnet (*Early Greek Philosophy*, 3rd edit., London, 1920, pp. 187-189) thinks, is not certain. If Burnet is right, the conception of the heavens is interesting, though still obscure. If one of the bands is the Milky Way, as seems probable, the chiastic arrangement is intelligible.

THE SPHERICITY OF THE EARTH

came a legendary figure,[161] to whom in the course of time one might ascribe a knowledge of almost everything. We might pass on without more ado, were it not that many works still largely used by scientists are wholly uncritical in regard to Pythagoras. Many recent works also otherwise admirable must be carefully scrutinized in this respect, such as those of Tannery and his followers, who have made large claims for Pythagoras and the Pythagoreans in the effort to reconstruct the history of the mathematical sciences.

If we date the activity of Pythagoras in the last quarter of the sixth century, which for our purpose is definite enough, there was no earlier Greek science except that of the Ionians, and there are sufficient indications that it was not unknown to him. In view of the uncertainty regarding his scientific notions we must scrutinize such statements as we have and judge them on the principle of probability. Now Diogenes Laërtius,[162] apparently following Favorinus, a rhetorician of the second century of our era, says, " Pythagoras was the first to call the world a *cosmos,* and the earth round ($στρογγύλη$) ; according to Theophrastus it was Parmenides; according to Zeno it was Hesiod." Obviously the statement is vague as to Parmenides and Hesiod; are they supposed to have used both terms, or only the latter? Pythagoras, however, is credited with using both. Regarding the term " cosmos " we need not now inquire further. But what is meant by attributing to Pythagoras the use of " round "? The context certainly implies that it is a question primarily of terms; but one may take it for granted that " round " in this late, uncritical text meant " spherical "; that Pythagoras regarded the earth as spherical is distinctly implied also in a statement of Aëtius [163] that he divided the earth into zones after

---
[161] See Isidore Levy, *Recherches sur les sources de la légende de Pythagore,* Paris, 1926.
[162] VIII, 48.
[163] III, 14, 1.

74 THE FRAME OF THE ANCIENT GREEK MAPS

the analogy of the cosmic sphere. This, however, signifies little, since, as we have seen, the doxographic tradition, influenced in its later stages by Stoic " accommodation," was utterly uncritical. The case is quite different, however, when one is dealing with Theophrastus; for, though he undoubtedly made many mistakes in interpretation, misled by the pronouncements of his teacher Aristotle, he was a man of intelligence and had studied the record. It is plain, then, that the text of Diogenes Laërtius does not justify us in thinking that his claim for Pythagoras is supported by the authority of Theophrastus, who is cited rather in favor of the claim of Parmenides. But, even so, the question arises: what did Theophrastus claim for Parmenides? Apparently only that he used a term meaning " round " in speaking of the earth. Since the term is itself ambiguous, being equally applicable to a circular disk and to a sphere, we gain nothing from the text either for Pythagoras or for Parmenides. It is certain that in the fifth century the term in question was not used exclusively or even generally with reference to a sphere.[164]

We have therefore to consider the probabilities in the case. To do so we must build on what we have reason to regard as certain. As we have seen, Aristotle tells us that Anaximenes, Anaxagoras, and Democritus explained the fact that the earth does not fall by its broad, flat shape. Whether or not these philosophers really had that problem in mind, Aristotle must be assumed to have known that they agreed with the earlier Ionians in regarding the earth as a flat disk. With Anaxagoras and Democritus we come to the latter part of the fifth century before Christ, and to their number we must add Archelaus, the disciple of Anaxagoras; for he also accepted the old Ionian view of

[164] See Erich Frank, *Plato und die sogenannten Pythagoreer*, Halle, 1923, pp. 199-200. Burnet, who insists that στρογγύλη must mean " spherical " in earlier philosophies, concedes (*op. cit.*, p. 357) that in connection with Diogenes of Apollonia (Diogenes Laërtius, IX, 57) the word is used to describe a disk.

THE SPHERICITY OF THE EARTH 75

the earth.[165] It is noteworthy that none of these philosophers appears from the records to have given any thought to the possibility that the earth is a sphere. This circumstance might be explained by the fact that they belonged to the eastern part of the Greek world, whereas Pythagoras and Parmenides were active in southern Italy. Unfortunately this suggestion is not satisfactory. Parmenides undoubtedly came to Athens while Anaxagoras resided there,[166] and he is said to have been at Thurii at or shortly after the founding of that colony in southern Italy (444-443 B. C.). Herodotus also shared in that venture and might have learned there something about the form of the earth, in which he was clearly much interested, either from Parmenides or from others who knew him. There is nothing in Herodotus, however, to suggest that he doubted the Ionian view except as regards the encircling River Oceanus. But in the cases of Anaxagoras and Democritus their silence and indifference, if they knew of an alternative view, would be inexplicable. Both were competent mathematicians, more especially Democritus, and they laid the mathematical foundations of perspective,[167] which has a direct bearing on the conception of the shape of the earth. We should expect them to be greatly excited by the suggestion that the earth was really spherical; and it is difficult to conceive that they would not have heard the suggestion if it had been made by Pythagoras or Parmenides, or even by their followers before the last quarter of the fifth century. As for Anaxagoras, it is generally, and in my opinion rightly, agreed that he had made a study of Parmenides and his successors, because his philosophy, like that of his contemporary Empedocles, can be understood only as an attempt to reconcile the fact of change with the principle of permanent being, so drastically enunciated by

[165] Hippolytus, *Refutatio*, I, 9, 4.
[166] The date is in dispute.
[167] Vitruvius, *De architectura*, VII, praefatio, 11.

Parmenides. The same is true of Democritus; and Aristotle definitely derives the atomic theory of Leucippus and Democritus from the Eleatic philosophy of Parmenides and Zeno.[168] Furthermore, later tradition represented Democritus as a Pythagorean,[169] although this suggestion means little, because that same uncritical tradition tended to claim every man of note as belonging to the school of Pythagoras.

The probabilities, therefore, are heavily against the assumption that either Pythagoras or Parmenides thought of the earth as a sphere. If we reject their claims in this matter, we must likewise refuse to believe that Parmenides taught the doctrine of the geographical zones,[170] in any sense, at least, other than that of the Ionian *oikumene* and its fringes, for the later concept of the zones was based on the projection of the celestial circles upon a spherical earth. We shall presently see that our earliest unambiguous and certain evidence points to the turn of the fifth and fourth centuries as the time when both of these conceptions were first suggested.

[168] Aristotle, *De generatione*, 324 b 35-325 a 25. Theophrastus, according to Simplicius (*Physica*, in Hermann Diels' edition in the *Commentaria in Aristotelem graeca*, Berlin, 1882-1909, Vol. 9, p. 28, line 4), seems (as Burnet, *op. cit.*, p. 332, says) to have thought that Leucippus was actually a pupil of Parmenides. Diogenes Laërtius (IX, 41) appears to say that Democritus mentioned Oenopides. If that were true, it would seem to follow that the latter, who was one of the leading astronomers toward the close of the fifth century, did not expound the doctrine of the sphericity of the earth.

[169] See Frank, *op. cit.*, p. 185, note 1.

[170] Strabo (II, 2, 2) reports: "Posidonius states that Parmenides was the originator of the five zones, but that he represented the torrid zone about double in width, extending beyond the tropics into the temperate zones." We shall presently see that Aristotle thought that the earth becomes uninhabitable by reason of the heat even before one reaches the tropics, but it is not likely that he owed the notion to Parmenides. Karl Reinhardt (*Parmenides und die Geschichte der griechischen Philosophie*, Bonn, 1916, p. 147, note 1) rejects this report of Posidonius, asserting that the terrestrial zones were unknown even to Plato. Erich Frank (*op. cit.*, pp. 198-200) suggests that it was Eudoxus, the associate of Plato, who first transferred the celestial zones to the earth and thus introduced the notion of the "clime" for the vague "regions" (τόποι) of earlier thinkers. See also Friedrich Gisinger, *Die Erdbeschreibung des Eudoxos von Knidos* (constituting *Stoicheia*, Vol. 6), Leipzig, 1921, p. 130.

THE CLAIMS OF ARCHELAUS

Before turning to this evidence we must briefly refer to an idea that has been considered as a first attempt to account for certain phenomena that find their explanation in the rotundity of the earth. Hippolytus, Bishop of Rome in the first quarter of the third century of our era, published an extract from the work of Theophrastus dealing with the opinions of the physical philosophers. This is regarded as uncommonly trustworthy, although Hippolytus did not himself make the epitome and it does not always accurately represent the original. The account that it contains of the doctrines of Archelaus, the disciple of Anaxagoras, is particularly sketchy and incomplete. It does not expressly describe the shape of the earth, but obviously assumes that it is a disk. The account as Hippolytus gives it reads [171]: " He [Archelaus] says that the heavens were inclined and that then the sun made light upon the earth, made the air transparent, and the earth dry; for it was originally a pond, being high at the circumference and hollow in the centre. He adduces as proof of this hollowness that the sun does not rise and set at the same time for all peoples, as it ought to do if the earth were level." This is, indeed, a remarkable statement. That the earth was originally in the same plane with the sun and was subsequently brought under its rays by a dip of the earth at the south was, as we have seen, common Ionian doctrine. No less so, we may be sure, was the assumption that the rim of the earth was raised, because it would otherwise be difficult to explain the River of Oceanus, which must have had banks high enough to hold in its waters despite the dip of the earth. That the Ionians assumed the existence of lofty mountains in the north we know, since they thus

---

[171] I, 9, 4. I quote the rendering of Burnet, *Early Greek Philosophy*, 3rd edit., London, 1920, p. 359.

accounted for the darkness of night.[172] There is every reason to think that the Ionians generally thus conceived the surface of the earth as somewhat hollowed at the center.[173] The difficulty arises when we are asked to believe that Archelaus confirmed the hollowness of the earth's surface by the consideration that the sun does not rise and set for all peoples at the same time, a statement upon which those rely who think that the cup shape of the earth was suggested to explain phenomena that are accounted for by its rotundity.[174] Living at Athens, he would of course know that as the sun rises the shadow of Mt. Hymettus grows shorter and gradually disappears; and he could conceive that a lofty range of mountains on the eastern rim of the earth would have a similar effect.[175] But it is obvious that the effect would be just the opposite of that produced by the rotundity of the earth; for the sun would rise later, instead of earlier, as one approached the eastern rim of the cup, and set earlier, rather than later, near the western rim. For the intermediate regions, which were best known to the Greeks, there were no data available that called for an explanation. We naturally suspect, therefore, that, though Archelaus may well have asserted that

[172] Hippolytus (I, 7, 6) reports this of Anaximenes; but it is certain that it was also common Ionic doctrine.
[173] If Anaximander (as Hippolytus, I, 6, 3 reports) compared the shape of the earth to the drum of a column, this would be true, as one can readily see by examining such drums, which, wherever they are found, are somewhat hollowed. I have long suspected that the text is corrupt and that Anaximander really referred to the upper millstone, which also was hollowed both above and below.
[174] See Frank, *op. cit.*, pp. 187-189.
[175] Aristotle (*Meteorology*, 350 a 30-34) says that the great height of the Caucasus is proved by the fact that it can be seen from the so-called " Deeps " of the Euxine and the entrance of the lake (Sea of Azov?), and that its summit is illuminated by the sun a third part of the night after sunset and again before dawn. There is obviously something wrong with the text; but even if the statement were true, these facts would not necessitate a change in one's conception of the earth as a whole. It seems certain that the estimates of the height of various mountains made by Aristotle's pupil Dicaearchus were made expressly to prove that, however high, the elevations did not appreciably alter the contour of the earth.

the earth was high at the circumference and hollow in the center, he did not conceive of its shape as explaining the phenomena in question.

Sphericity of Earth Probably Unknown Before End of Fifth Century Before Christ

Thus a critical examination of the evidence lends no support to the claims of early Greek thinkers to the honor of having suggested the sphericity of the earth. The doxographic tradition, as it appears in Aëtius and Diogenes Laërtius, does indeed attribute the discovery to Pythagoras and Parmenides, but it is doubtless significant that there is no unambiguous evidence that Theophrastus, from whose historical treatise the doxographic tradition ultimately derives, claimed the honor for either of these worthies. At best we might cite him in favor of Parmenides and against Pythagoras, which would impugn the trustworthiness of the later doxographers; but even that cannot be granted, because strictly we have a right to infer no more than that Parmenides used an ambiguous word meaning " round " with reference to the earth. The later doxographic record, which alone expressly attributes the discovery of the sphericity of the earth to thinkers earlier than the fourth century, is not only discredited in part by Theophrastus, but it reveals the influence of the Stoics, who were notoriously uncritical in such matters. In particular we have reason to suspect that Posidonius and his school, in the first century before Christ, contaminated the record.

For the period before Socrates, then, the result of our study is negative. Not only is there no positive evidence that any one had suggested the sphericity of the earth, but the reasonable inference from what we know supports the negative conclusion. In the earlier time we might assume a want of ready communication between east and west as explaining the failure of the Ionians to take account of discoveries made in southern Italy, though this is hard to maintain in view of the relation existing between the thought of Parmenides and Heraclitus; but from the middle of the fifth century onward, when east and west met at Athens, this excuse is futile. Athens, alive to everything new and keenly debating every question, knew what all the world was thinking. There, if anywhere, one could have learned of the great discovery, had it been made. That Anaxagoras and Archelaus apparently had no notion of the sphericity of the earth is surely important evidence that the suggestion had not been offered.

This negative result needs to be emphasized because Berger not only credited Parmenides with a knowledge of the sphericity of the earth and the five geographical zones but even assumed that the latter were known to Xenophanes at the end of the sixth century.[176] Even Burnet, who in general saw clearly the lines of historical development, rather curiously concluded that Pythagoras held the earth to be a sphere [177]; and many writers have subscribed to the views of Berger and Burnet.

[176] See Hugo Berger, *Geschichte der wissenschaftichen Erdkunde der Griechen*, 2nd edit., Leipzig, 1903, pp. 207-209 and *passim*.

[177] *Early Greek Philosophy*, 3rd edit., London, 1920, p. 190, note 1. Hippolytus (I, 7, 5) reports that Anaximenes assumed the existence of unseen earthy bodies borne along with the stars. The statement is repeated by Aëtius (II, 13, 10). This presumably referred to meteors, as is clearly stated in connection with Diogenes of Apollonia (Aëtius, II, 13, 9), who harked back to the theories of Anaximenes. Burnet, however, assumes that Anaximenes meant by this assumption to account for eclipses. Eudemus is said (Theo Smyrnaeus, *On the Mathematical Knowledge Which is Needed to Read Plato*, Eduard Hiller's edit., Leipzig, 1878, p. 198, line 16) to have attributed the explanation of eclipses to Anaximenes; but there is obviously confusion here, because Anaximenes thought the moon was made of fire. If the moon was conceived to be fiery, its phases were obviously not explained by its relation to the sun. Lunar eclipses also would have to be otherwise explained, and solar eclipses likewise. One suspects that Eudemus referred not to Anaximenes, but to Anaxagoras; for Plutarch (*Nicias*, 23) was doubtless right in attributing the discovery of the true explanation of eclipses to Anaxagoras. The confusion may again be due to Posidonius (see above pp. 66-67). In a later article in *Scientia* (" L'expérimentation et l'observation dans la science grecque," Vol. 33, 1923, p. 95) Burnet flatly asserted that Pythagoras discovered the sphericity of the earth and said that it was doubtless suggested by the observation of lunar eclipses!

CHAPTER VII

PLATO AND ARISTOTLE AND THEIR SOURCES

PLATO ON THE SPHERICITY OF THE EARTH

The first unequivocal evidence for the conception of the earth as a sphere is found in Plato. In the *Phaedo*[178] Socrates is made to say that he once heard some one reading from a book stated to be the work of Anaxagoras. Hearing that Intelligence was therein asserted to rule the world, Socrates conceived great hopes of learning the nature of things, because Intelligence must of course order all things for the best. One of the questions on which he would fain be enlightened was whether the earth was flat or round. His ardent hopes were dashed, however, because the book did not explain the world teleologically, the causes alleged being purely mechanical.

In this instance we may be sure that the word rendered " round " really means spherical. Aside from the complaint of Socrates about the absence of a teleological explanation, we may confidently assert that he could have learned nothing about a spherical earth from Anaxagoras because, as we have seen, that philosopher still regarded the earth as flat. Nevertheless it is obvious that someone had suggested that it was not flat but round. Farther on in the same dialogue[179] Socrates expresses his belief that the earth is really round and at the center of the heavens, needing neither air nor any other thing to support it. Socrates, indeed, asserts his belief[180] that, as somebody has per-

---
[178] 97 b-99 d, 110 b.
[179] 108 e-109 a. See *Timaeus,* 63 a, and A. E. Taylor's note *ad loc.* in his *Commentary on Plato's Timaeus,* Oxford, 1928.
[180] *Phaedo,* 108 c.

82 THE FRAME OF THE ANCIENT GREEK MAPS

suaded him, the earth is neither of the character nor of the size that those who are wont to discourse about it suppose. To whom he owed the notion he does not say, and we have no means of determining. It is important, however, to note that Socrates does not pretend to have knowledge of the truth, and the impression made by his cautious statement is confirmed by his comments on the book of Anaxagoras. Since the description that Socrates gave of the character and size of the earth differed from that of those who were wont to discourse about the earth and who must have been nearly or quite his contemporaries, we infer that his view was quite new and that it was not the accepted view of geographers. This is, indeed, the natural conclusion to be drawn from everything that we know; for the arguments for the sphericity of the earth were not suggested by a study of the earth itself, but by mathematics and astronomy.

As to the date when the hypothesis was proposed, we have no knowledge and must rely on inferences from rather vague data. Socrates mentions Anaxagoras, by whom he was not enlightened. This presumably refers to a time when Anaxagoras was no longer at Athens; otherwise it would not have been difficult to refer to the philosopher himself rather than to his book, for Socrates surely was not slow to ask questions of any man he met. Now Anaxagoras fled from Athens when he was accused of impiety; but there is a difficulty about the time of that event, as the ancient accounts do not agree. Burnet[181] concludes that it was in 450 B. C., while others place it about 431. I incline to the later date. At all events, we are concerned with ideas of the latter half of the fifth century. We may assume that Archelaus lived on until about 430; and if, as seems probable, Aristophanes in the *Clouds*,[182] first per-

[181] John Burnet, *Early Greek Philosophy*, 3rd edit., London, 1920, p. 256.
[182] 264.

formed in 423, was satirizing the views of Diogenes of Apollonia, we may conclude that this late Ionian was still proclaiming a flat earth supported by air. We thus narrow to a quarter of a century the interval between the last adherent of the flat earth and the dramatic date of the *Phaedo* (399). This, of course, assumes that the *Phaedo* reflects the opinions of Socrates rather than of Plato, an assumption vigorously supported by Burnet and A. E. Taylor but as vigorously contradicted by other scholars. Fortunately, since we have sufficiently limited the interval, we need not enter here into this debate, in view of the impossibility, which is generally admitted, of naming the author of the hypothesis.

Though we cannot satisfy the natural desire to know who first suggested a view so important, we are not prevented from drawing certain inferences as to the quarters from which the hypothesis emanated. It is more than reasonable to believe that we should not seek the author among the Ionians and their adherents, since to the very last they seem to have held fast to belief in the flat earth. Furthermore, the geographers who " were wont to discourse about the earth " were obviously not in agreement with the views expressed by Socrates as to the character and size of our planet. Where, then, should we seek the unknown by whom Socrates (or Plato) was persuaded that the earth was a sphere? As we have already remarked, the arguments adduced to support the hypothesis are astronomical and geometrical rather than geographical. Geography was peculiarly the creation of the Ionians and certainly one of their major interests. It happens that I do not at all agree with those who attribute the entire early development of mathematics to Pythagoreans, but certainly they were more concerned with this subject than were the Ionians, and there is really no trace of contributions by Pythagoreans to geography as it was understood in the fifth century. The presumption, therefore, is that the hypothesis originated among those who followed the school of Pythagoras, and, since it is now generally agreed that later tradition fathered upon Pythagoras himself discoveries made by his followers, this presumption is somewhat strengthened by the fact that the hypothesis of the sphericity of the earth was attributed to Pythagoras and to Parmenides, who was accounted a Pythagorean. Finally we have to take account of the fact that in the *Timaeus* [183] of Plato also the sphericity of the earth and the possibility, at least, of antipodes are taken for granted. Since Timaeus himself was unquestionably a Pythagorean, we seem to receive from Plato a definite hint as to the source of the suggestion that Socrates in the *Phaedo* accepted without asserting that he could prove it to be true.

Socrates himself, besides having among his intimates during his later years the Pythagoreans Simmias, Cebes, and Phaedondas, was evidently on intimate terms with mathematicians [184] who, though not themselves Pythagoreans, must have been in touch with the mathematical studies of the Pythagoreans; and Plato is acknowledged to have cultivated intimacy with Archytas of Tarentum, one of the latest representatives of the old Pythagorean school. Thus, without regarding Socrates and Plato as Pythagoreans, we have good reason for thinking that either of them might well have owed the suggestion of the sphericity of the earth to some adherent of that school, and I do not hesitate to accept this conclusion. But as between Socrates and Plato it would be hazardous to decide who first accepted the suggestion. At all events, the date can be reasonably fixed between 425 and 375 B. C., and the vagueness of the description of the earth given in the *Phaedo* [185] shows that the conception had not gone far

[183] 63 a.
[184] Theodorus of Cyrene and Theaetetus of Athens.
[185] 108 d-113 c. Burnet, in his annotated edition of the *Phaedo,* Oxford, 1911, recognized the eclectic character of the picture of the earth and suggested
(continued on next page)

PLATO AND ARISTOTLE 85

beyond the tentative stage, in which elements of quite different origin were promiscuously brought together. It is noteworthy that Plato shows no knowledge of geographical zones; and his statement [186] that the earth as seen from above would look like a varicolored ball made of twelve pieces of leather (alluding, as it presumably does, to the inscription of the dodecahedron in the sphere) suggests a purely geometrical construction having no relation to the traverse of the sun.

ARISTOTLE ON THE SPHERICITY OF THE EARTH

Aristotle is the first author who affords us clear evidence bearing on the geography of the spherical earth. As has already been pointed out, he is the first to reveal an appreciation of the significance of the position of the sun. " There are," he says,[187] " two inhabitable sections of the earth: one near our upper or northern pole, the other near the other or southern pole; and their shape is like that of a tambourine. If you draw lines from the centre of the earth they cut out a drum-shaped figure. The lines form two cones; the base of one is the tropic, of the other the

Continuation of footnote 185.
that it may have been the conception of Socrates himself. Max Wellmann, " Eine pythagoreische Urkunde des IV. Jahrhunderts v. Chr.," in *Hermes*, Vol. 54, 1919, p. 243, thought that *Phaedo*, 109 a, referred to the author of the Pythagorean excerpt from Alexander Polyhistor given by Diogenes Laërtius (VIII, 24-33). Diels seems to have accepted Wellmann's view. Nevertheless it appears to me very doubtful because of the extremely eclectic nature of the ideas expressed in the excerpt, which Wellmann himself recognized. If Plato's source were really Pythagorean, which Burnet denied, it would prove only that the doctrine of the sphericity of the earth had as yet led to no favorable change in geography.
[186] *Phaedo*, 110 b.
[187] *Meteorology*, 362 a 32-b 9 (E. W. Webster's translation, in *The Works of Aristotle Translated into English*, Oxford, Clarendon Press, 1923). Questions relating to the shadow cast by the sun are discussed in the Pseudo-Aristotelian *Problems* (XV, 5 and 9), which are probably due to the school of Strato, who succeeded Theophrastus in the headship of the Lyceum early in the third century before Christ. See also Aristotle, *Liber de inundacione Nili*, fr. 248, in Valentin Rose, *Aristotelis qui feruntur librorum fragmenta*, Leipzig, 1886.

ever-visible circle, their vertex is at the centre of the earth. Two other cones towards the south pole give corresponding segments of the earth. These sections alone are habitable. Beyond the tropics no one can live; for there the shade would not fall to the north, whereas the earth is known to be uninhabitable before the sun is in the zenith or the shade is thrown to the south; and the regions below the Bear are uninhabitable because of the cold." These words show that we are dealing with a purely mathematical construction. As such there is no fault to be found with it; but when we consider its application to the earth as we know it, we see that no dependable data were available. Aristotle knew that south of the northern tropic, the shadow of the sun would shift, but he had no notion of the location of the tropic on the earth's surface. As for the southern tropic, it was obviously merely inferred as a consequence of the shape of the earth and its relation to the sun. That this was essentially a mathematical construction of astronomers is not only in itself obvious but is confirmed by several statements of Aristotle himself. Thus, in support of his contention that the earth is at the center of the universe, both in fact and by reason of its very nature, he says [188]: "This is confirmed by what the mathematicians say regarding astronomy; for, though the figures change whereby the arrangement of the stars is defined, their appearance is such as follows from the fact that the earth lies at the center." Here, as in the purely theoretical description of the zones, we have the clearest evidence that we are dealing with exact studies based on spherics, a branch of mathematics that was just beginning to be cultivated about 400 B. C. Similarly Aristotle refutes [189] the objection urged against the sphericity of the earth that at the rising and setting of the sun the section cut by the line

[188] *De caelo*, 297 a 2-6.
[189] *De caelo*, 293 b 34-294 a 8.

of the horizon is not curved but straight, by referring to the facts that the distance from the earth to the sun is great and that the arc of the earth's curve, which covers a portion of the sun's disk, being very small in relation to the size of the globe, would necessarily appear straight. Here he is plainly relying upon mathematical data supplied by the study of perspective and of the properties of the sphere.

Aristotle is fully convinced of the sphericity of the earth. He says [190] that the earth lies at the center of the universe because it (and every particle of it) tends naturally to the center. Assuming an origin of the earth, the particles, tending equally from all sides toward the center, would inevitably form a sphere, and consequently the earth itself must be a sphere. The same form would result from the pressure of greater masses on smaller ones, even if the particles coming from without were not added equally from all sides, since all the stress of weight is directed toward the center. Since Aristotle himself did not regard the earth as originating, it is clear that this argument for him signified little. Of more consequence, no doubt, was the consideration that he adds, to wit, that all falling bodies tend to the center and do not describe in their fall parallel lines but lines converging at the center. Doubtless this also was an inference drawn from the study of the sphere, since he does not include it among the evidence that he adduces from observation. "Furthermore," he says,[191] "the sphericity of the earth is proved by the evidence of our senses, for otherwise lunar eclipses would not take such forms; for whereas in the monthly phases of the moon the segments are of all sorts—straight, gibbous, and crescent—in eclipses the dividing line is always rounded. Consequently, if the eclipse is due to the inter-

---
[190] *De caelo*, 297 a 8-b 21.
[191] *De caelo*, 297 b 23-298 a 20.

position of the earth, the rounded line results from its spherical shape. Besides, by the appearance of the stars it is made evident not only that the earth is round, but that it is of no great size; for if we change our location even a little to south or north the horizon manifestly changes,[192] so that the stars vertically overhead change considerably and are not the same as we move to south and north. Indeed, certain stars are visible in Egypt and about Cyprus but are not seen in northern regions; and stars which farther north are always visible are known to set in the former regions. Hence it is evident not only that the earth is round in shape but also that its sphere is not large; for otherwise the change would not be so marked when one moves so short a distance. Consequently, it is thought that those do not assume what is incredible who suppose that the region about the Pillars of Herakles is adjacent to that about India and that the [intervening] sea is one [193] [and the same]; and in support of this they refer to the fact that elephants [194] are to be found in both these regions, assuming that their occurrence there is due to the proximity of these extremes to each other. And those of the mathematicians who attempt to calculate the circumference say that it is about 400,000 stades. From such indications we must conclude not only that the mass of the earth is

---

[192] See *Meteorology*, 365 a 29-31.

[193] From the Aristotelian *Liber de inundacione Nili* (fr. 248, in Valentin Rose, *Aristotelis qui feruntur librorum fragmenta*, Leipzig, 1886) we learn that the ocean was asserted to be one by a certain Athenagoras, who appears to have flourished about 357-349 before Christ.

[194] The reference to elephants is curious. They were known both in India and in Africa in the fifth century, but that of itself signifies little. We should like to know who inferred from the presence of elephants in those regions their possible proximity. It was thought in antiquity that there was, or had been, a land bridge between India and Africa, and it is very probable that Plato used this conception as the basis of his fiction of the lost Atlantis in his *Critias*. It is therefore interesting to note that he mentions the existence of elephants in Atlantis (*Critias*, 115 b). There is apparently some connection between these references. See my "Suggestion Concerning Plato's Atlantis," *Proc. Amer. Acad. of Arts and Sci.*, Vol. 68, 1933, pp. 204-215.

spherical but also that it is not large in comparison with the size of the stars."

It is hardly necessary to comment at length on this passage. The repeated insistence on the smallness of the earth seems hard to explain except as an answer to Plato, who had suggested that it was much larger than geographers usually thought. If that be the correct explanation, it would follow that the attempt in question to estimate the circumference of the earth must have been made by mathematicians practically contemporary with Aristotle. This estimate, however, was still very much too large, as is seen by comparison with the much closer approximation of Eratosthenes, who gave 252,000 stades as the result of his calculations. Obviously the different results were due not only to different data but also to differences in method. While Eratosthenes measured the arc by taking the position of the sun at the summer solstice, we may infer that the calculation reported by Aristotle was based on observation of certain stars, more probably of groups of stars or constellations. The reference to stars seen in Egypt and about Cyprus, but invisible farther north, points, perhaps, to Eudoxus as the author of Aristotle's estimate; for, as we shall presently see, Eudoxus is reported to have made such observations on the star Canopus. His was apparently a first attempt.

All the indications, therefore, suggest a very rapid advance in mathematical geography, beginning about 400 B. C. At first it was pure mathematics applied to the study of the sphere; but once the essential geometrical problems were solved, certain consequences were seen to follow, which were found to accord with observed or observable facts. As yet these latter were few in number: data for the construction of maps were not yet at hand, and even Aristotle could make the mistake of believing that the earth became uninhabitable even before one reached the tropics. Clearly no attempt had yet been made to draw a map based on the new conception. After describing the zones as a purely mathematical construction, Aristotle repeated the old taunt that Herodotus directed at the maps of the Ionians, saying [195] that men *still drew* their maps ludicrously, representing the *oikumene* as a circle, which is contrary to observable fact no less than to reason. The state of knowledge touching these matters is clearly revealed in what he immediately proceeds to say: "If we reflect, we see that the inhabited region is limited in breadth, while the climate admits of its extending all round the earth. For we meet with no excessive heat or cold in the direction of its length but only in that of its breadth; so that there is nothing to prevent our traveling round the earth unless the extent of the sea presents an obstacle anywhere. The records of journeys by sea and land bear this out. They make the length far greater than the breadth. If we compute these voyages and journeys, the distance from the Pillars of Herakles to India exceeds that from Ethiopia to Maeotis and the outermost Scythians by a ratio of more than 5 to 3, as far as such matters admit of accurate statement. Yet we know the whole breadth of the *oikumene* up to the uninhabited parts: in one direction no one lives because of the cold, in the other because of the heat." There is here an interesting combination of the new suggestions based upon theory with the empirical facts revealed by the explorations of travelers and embodied in the Ionian maps.

In Plato and Aristotle, then, we find unmistakable evidence of a theoretical view of the spherical earth as based on pure mathematics and a study of astronomy; but we see also that the observations necessary for the construction of a map were still wanting. Although a point of view had been attained that enabled one to attempt an estimate of

[195] *Meteorology*, 362 b 12-27.

the circumference of the earth, the method was not yet precisely formulated, and the data, astronomical as well as geographical, were still too inaccurate to lead to satisfactory results. Everything, therefore, points to the conclusion that the problems connected with mathematical geography were just being broached for the first time and in a tentative way. This accords also with what we know of the history of mathematics and astronomy. As for the maps in current use, there is no possible doubt that they were of the type made by the early Ionians and that, despite any improvements that may have been introduced in detail, they did not much depart from the form they received at the hands of Anaximander and Hecataeus.

We cannot but wonder, therefore, how Berger and other scholars, whose merits in other respects we gladly acknowledge, could accept the uncritical reports of later Greeks who attributed the conception of the spherical earth to Pythagoras or Parmenides, in the sixth and early fifth centuries, and the discovery of the zones to Xenophanes, in the sixth century, or even to Parmenides. The reference of the latter discovery to Parmenides, we know, was made by Posidonius—on what grounds we do not know, but we readily surmise that Posidonius had no other evidence than the so-called " crowns " mentioned in the cosmography of Parmenides. These were nothing more than celestial bands, like the " whorls " described by Plato in the *Republic,* and were related to the series of circles conceived by Anaximander. Whether Posidonius merely mistook them for terrestrial bands or assumed that Parmenides must have meant by his " crowns " the celestial zones and transferred them, as a matter of course, to the spherical earth, as later astronomers did, we cannot say, and it is idle to speculate; but we are justified in holding, with the most competent students of these problems, that Posidonius was surely mistaken.

Possible Origins of the Concepts of Plato and Aristotle

In rejecting the claims of Parmenides and still earlier philosophers we seem to be on solid ground; but it is quite evident that Plato and Aristotle were not themselves responsible for the new views. Plato makes Socrates say that someone has persuaded him that the character and size of the earth are quite different from the notions on the subject entertained by geographers, and this presumably implies the sphericity of the earth as well as its large extent. Plato, however, does not tell us who suggested these newer views. Aristotle also is very indefinite, naming no authority for the views that he accepts and referring only to the mathematicians who estimate the circumference of the earth. It is noteworthy that both Plato and Aristotle habitually use the present tense in referring to these matters, thus creating the presumption that they are dealing with current opinions. In the case of Plato the natural inference is that he (or Socrates) owed the suggestion to someone who stood in some relation to the Pythagoreans; and Aristotle conducts his discussions in a way to suggest that he is chiefly concerned with notions debated in the school of Plato. This state of affairs tends to limit the field of inquiry when we ask by whom these discoveries were made.

Berger makes much of a passage in the *Clouds*[196] of Aristophanes, first performed in 423 B. C., where Strepsiades is represented as seeing in the "Worry Hall" of Socrates a map of the world and instruments called respectively "astronomy" and "geometry." "Astronomy" was undoubtedly some sort of instrument for astronomical purposes, but precisely what does not now concern us. As for the map, we now know approximately how it was

---

[196] *Clouds*, 201-207.

constructed. But how we should picture "geometry," represented as an instrument to be used "in measuring the whole earth," is hard to imagine. Why Berger should insist that it was a globe we cannot comprehend, for Strepsiades asks whether the instrument is to be used in surveying lands allotted to colonists. Since actual objects displayed by the actors are not necessarily presupposed, the thought of measuring the whole earth is the only thing that is important. But here, again, we are left in the dark, because there was always a possible confusion between "earth" as a whole and "earth" in the sense of *oikumene* or land mass. While, therefore, it is possible that Aristophanes was referring to some attempt of a contemporary to determine the actual size of the earth as a whole, it is equally possible that he had in mind merely a surveyor's cord or "chain," with which one could measure distances on the earth and in that way ascertain the shape and the relative length and breadth of the *oikumene,* a matter that was being debated at the time. Berger has here, as in other instances, too hastily and uncritically read his own meaning into the ancient record. While the possibility must be conceded, there is in the *Clouds* no certain evidence that the earth had then been conceived as a sphere. Even if that possibility were accepted as a fact, we should still be at a loss to name the originator of the conception.

Philolaus

In considering who among the Pythagoreans first made the suggestion that the earth was a sphere we naturally think of Philolaus; for he was a contemporary of Socrates and known to him. That he conceived the earth as spherical is not definitely stated, but it might be inferred from the report of the doxographers that he did not locate the earth at the center of the universe but made it circle round the central fire in an oblique orbit like the sun and moon [197]; the earth would thus be regarded as a planet and doubtless as spherical. Unfortunately there is the gravest doubt concerning the genuineness and the date of the philosophical work attributed to Philolaus. Despite the arguments that have been advanced in support of its authenticity I am still convinced that it dates at the earliest about a century after Socrates. Burnet, who rejected as spurious the philosophical fragments of the work attributed to Philolaus, thought nevertheless that the description of the earth given by Plato in the *Phaedo* might have been inspired by Philolaus.[198] This, however, is a mere conjecture, without any foundation in the record.

Archytas

If we disregard Philolaus as too uncertain and as at best only a very shadowy figure, we naturally think of another Pythagorean, Archytas of Tarentum, because of Plato's personal relations to him; but, again, we unfortunately know too little about his opinions in the matters here in question. It is altogether probable, indeed, that he held the earth to be spherical, and he may have formed some notion of its size. Horace [199] calls him a "measurer of earth and sea and the numberless sands," which might imply that he had somehow computed not only the surface area but also the volume of the earth and consequently that he had calculated its circumference. There are, therefore, some eminent scholars who regard him as the source of Plato's views about the earth. We know that he was a mathematician and mechanician of note, and the poet's statement that he "tempted the aërial abodes" may possibly

---

[197] Aëtius, III, 13, 2.
[198] John Burnet, "L'expérimentation et l'observation dans la science grecque," in *Scientia*, Vol. 33, 1923, p. 95.
[199] *Odes*, I, 18.

allude to his contrivance of a mechanical dove capable of flight; but confirmatory evidence of the characterization of Horace is entirely wanting.[200]

This fact raises the question whether the poet is to be taken literally at his word or whether he may not have used the license accorded to poets by adding traits derived from other men of similar interests, as Aristophanes in the *Clouds* borrowed traits of various philosophers and Sophists in characterizing Socrates. That the latter alternative needs to be considered is emphasized by the circumstance that Archytas is described as a measurer of the numberless sands; for one thinks inevitably of Archimedes and his *Sand Reckoner,* a work in which he developed a system capable of expressing numbers far greater than would be required to count the grains of sand contained in a sphere of the earth's size. If Archytas had anticipated Archimedes in this matter, for which the latter was celebrated, is it quite conceivable that we should hear of Archytas' feat only in a passing allusion by a Roman poet? We shall have to content ourselves, then, with saying that we do not know how much Plato may have owed to Archytas.

Eudoxus of Cnidus

As for Aristotle, his discussion moves entirely in the circle of interests that occupied the attention of the Academy, where the questions raised by Plato were of course debated. Of the opinions of the other members of Plato's school we know too little to be of help to us; but we cannot disregard the possibility that Eudoxus of Cnidus made important contributions to the questions at issue.[201]

[200] Unless one should find it somehow in Plato's *Theaetetus,* 173; but that is too general and too vague.
[201] See Paul Tannery, *Recherches sur l'histoire de l'astronomie ancienne,* Paris, 1893, pp. 14ff.; Hugo Berger, *Geschichte der wissenschaftlichen Erdkunde der Griechen,* 2nd edit., Leipzig, 1903, pp. 246ff., 264ff.

96 THE FRAME OF THE ANCIENT GREEK MAPS

He was one of the greatest mathematicians and astronomers of Greece, and he had been associated with Plato in the Academy. Whether or not Aristotle had known him personally, other members of the school, with whom Aristotle consorted, knew him and shared his interests. Aristotle repeatedly refers to his opinions on other matters. Now, as we have pointed out, Aristotle in discussing the shape and size of the earth mentions the fact that stars observed in Egypt and Cyprus are not visible farther north; and we know from other sources that Eudoxus made observations in Egypt and placed the star Canopus on the (movable) Antarctic Circle, i. e. the circle about the south celestial pole that corresponds to the horizon of Greece.[202] Another report has it that Eudoxus actually observed at Cnidus this same star, which he had seen in Egypt, but just above the horizon.[203] Such observations, given a knowledge of the mathematical principles involved, might well afford the basis for an estimate of the circumference of the earth. Moreover, as we have seen, Aristotle reveals the fact that in his time it was known that beyond the tropic the shadow shifts, though he gives no evidence that he knew either the location of the geographical tropic or the latitude of any locality as determined by the altitude of the sun at the meridian.[204] From this and from his statement that the earth becomes uninhabitable before one approaches the tropic, it is evident that he was not aware

[202] Hipparchus, *In Arati et Eudoxi Phaenomena commentariorum libri tres*, edited by Carl Manitius, Leipzig, 1894, pp. 114-116. See Sir Thomas Heath, *Aristarchus of Samos* . . . , Oxford, 1913, p. 192, note 2.
[203] Strabo, II, 5, 14.
[204] The reference to the declination of the Crown in the *Meteorology*, (362 b 9f.) is interesting and (unless interpolated, as Webster thinks; see note 187) possibly of the greatest importance for the history of scientific geography. Unfortunately it is extraordinarily vague; for we are told that, the region under the Bear being uninhabitable by reason of the cold, the Crown passes over "this region" and is over our heads when it stands in the meridian. Does "this region" refer to the frozen region or to the region of Greece? See Karl Müllenhoff, *Deutsche Altertumskunde*, Berlin, 1887-1900, Vol. 1, p. 235, note. I am strongly inclined to take Webster's view.

PLATO AND ARISTOTLE

that Syene lies approximately on the tropic itself, one of the data used by Eratosthenes in estimating the circumference of the earth.

We do not know from whom Aristotle derived the important data [205] that might have been used in arriving at the estimate of the earth's circumference that he reports as made by mathematicians of his day. While this is quite true, we have seen that observations attributed to Eudoxus, indefinite as they are, might well have furnished the basis for a calculation, assuming that he estimated somehow the distance between, say, Memphis and Cnidus and took account of the longitude of these places, for which of course he had no accurate data. Müllenhoff quotes [206] with approval the statement of Ideler that Eudoxus was merely an observing, not a measuring or calculating, astronomer; but we hesitate to accept this dictum in view of the Eudoxian system, which must rest on fairly accurate observations, checked by the astronomer, if he did not actually make them. Berger goes to the opposite extreme and asserts that Eudoxus must have determined the latitude of many localities, but Berger relied too much, perhaps, on the *Art of Eudoxus,* a treatise of much later date, the data of which it is not safe to attribute to Eudoxus himself without confirmation from other sources.

There is, however, a passage in the commentary of Hipparchus on the *Phaenomena* [207] of Aratus that is by some regarded as decisive. It may be translated as follows: " In the first place, Aratus seems to me to be mistaken in thinking the latitude of Greek lands to be such that the ratio of the longest day to the shortest day is as 5 to 3; for he says of the summer tropic, ' If you measure it as accurately as possible and divide it into eight parts, five

[205] See Berger, *op. cit.,* pp. 265f.
[206] Müllenhoff, *op. cit.,* Vol. 1, p. 240.
[207] I, 3, 5-7 (Manitius' edit., p. 26, 3ff.).

in the daylight will turn above the earth, and three below it.' Now it is agreed that in Greek lands the gnomon at the equinox is to its midday shadow in the ratio of 4 to 3. Consequently the longest day has a length of $14\frac{3}{5}$ hours and the latitude is approximately 37°. Where, however, the longest day is to the shortest as 5 to 3, the longest day has 15 hours and the latitude is approximately 41°. Consequently it is evident that this latter ratio does not hold for Greek lands, but rather for the region about the Hellespont. To be sure, Aratus in this matter does not present his own determination, but in this, as in others, he follows Eudoxus. But even if he did state this on his own authority without making clear to what locality he gave the said latitude, perhaps he should not be blamed on that account." On what grounds the omission of Aratus might be excused is left to be inferred; perhaps the reason was that he was writing a poem, in which too great accuracy might not be demanded.

The question of primary importance for us is, however, what inference we may safely draw from this passage regarding Eudoxus, whom, according to Hipparchus, Aratus was following. The one certain conclusion is that Eudoxus gave the ratio of the longest to the shortest day as 5 to 3. Did he also state the region for which that ratio holds? Hipparchus points out that it would be correct for the region of the Hellespont; and we are elsewhere told [208] that Eudoxus after his return from Egypt sojourned and taught at Cyzicus and at the Propontis. We may, then, possibly assume that the datum was determined there. But how much more can one gather from this discussion? Manitius [209] and Sir Thomas Heath [210] attribute to Eudoxus the entire method that Hipparchus indicates of determin-

---

[208] Diogenes Laërtius, VIII, 87.
[209] Pp. 293-294 of his edition of the *Phaenomena* of Hipparchus.
[210] *Aristarchus of Samos* . . . , Oxford, 1913, pp. 192-195.

ing latitude by the length of the longest day at the equinox and the solstice and vice versa. I cannot help thinking that we have no warrant for doing so. Eudoxus may have determined the ratio at the Hellespont of the longest to the shortest day at the solstices by empirical observation, without fully realizing the consequences for latitude.[211] Indeed, we may even say that this possibility is raised to a probability, when we consider that Aratus, who merely followed Eudoxus, accepted the datum as holding for Greek lands generally. This he would hardly have done if Eudoxus had specified, say, Cyzicus as the place of observation. Furthermore, Hipparchus offers a possible excuse for Aratus, which, if based upon the consideration that he was writing a poem, would not hold equally well for Eudoxus. One who reads the commentary of Hipparchus carefully will note other instances of rather vague statements of Eudoxus, whose locations of stars were far from being as accurate as those that Hipparchus himself made two centuries later. I conclude, therefore, that the detailed method described by Hipparchus should not be attributed to Eudoxus, but that it is to be regarded as due to Hipparchus, who uses it to point out the inaccuracy of Aratus and Eudoxus.

Eudoxus is credited with a map of the earth and a general treatise on geography. The former we must apparently accept, as the report concerning it [212] seems to go back to Eratosthenes' sketch of the history of geography; but of its character we have no satisfactory evidence. Regarding the geographical treatise, though its existence

[211] It will be recalled that Pytheas also, who is thought to have been a pupil of Eudoxus, did not determine the latitude of Massalia directly. Berger (*op. cit.*, p. 341) insists that the calculation is due to Hipparchus or possibly to Eratosthenes; Pytheas merely gave the empirical datum of the ratio of the shadow to the height of the stylus or gnomon. The determination of latitude in degrees is almost certainly the work of Hipparchus.

[212] Agathemerus, I, 1 (Carl Müller, *Geographi graeci minores*, Paris, 1882, Vol. 2, p. 471).

need not be denied, we cannot speak with any assurance [213]; for I have long had, and still have, the gravest doubts regarding many of the fragments generally assigned to it. It seems more probable that they derive from a work of later date, written by another Eudoxus called a "historian," or from one erroneously attributed to the Cnidian. At all events, they yield nothing of importance for scientific geography.

Berger [214] says that Eudoxus must have accepted the view that the earth is a sphere, because he estimated the length of the *oikumene* at twice its breadth; but this conclusion is wholly unwarranted, for the authority who makes this statement about Eudoxus [215] likewise reports that Democritus was the first to recognize that the earth (meaning, of course, the *oikumene*) was oblong, with a length half again as great as its breadth. Since Democritus is acknowledged to have regarded the earth as a flat disk, we cannot well infer that because an estimate is given of the length and breadth of the *oikumene* it must necessarily refer to a spherical earth. We may doubtless accept the view that Eudoxus held the earth to be a sphere, but our faith must rest on other grounds.

One more datum calls for consideration. The doxographic tradition [216] reports that, according to Eudoxus, the Egyptian priests explained the Nile floods by the alternation of the seasons: "for when we under the summer tropic have summer, those who live under the winter tropic have winter, and it is from among them that the flood waters rush down in torrents." We need not be detained

---

[213] The treatment by Friedrich Gisinger, *Die Erdbeschreibung des Eudoxos von Knidos*, constituting *Stoicheia*, Vol. 6, Leipzig, 1921, is in my opinion untrustworthy.
[214] *Op. cit.*, p. 247.
[215] Agathemerus, I, 1, 2.
[216] Aëtius, IV, 1. 7.

PLATO AND ARISTOTLE 101

by the supposed authority of the Egyptian priests,[217] to whom it was the fashion to attribute the discoveries of Greek science, for there is no reason whatever to believe that they had any notion of a spherical earth or of the existence and location of the tropics, which the report implies. But are we justified in accepting the statement even as a fiction fathered on the Egyptian priests by Eudoxus of Cnidus? Plato, as we have remarked, conceived at least the possibility of antipodes, and a Hippocratic treatise,[218] which may be dated about 400 B. C. or a little later, says that the south wind, coming from the south pole, is naturally cold but becomes heated in passing over Libya. Thus the existence of northern and southern hemispheres, with similar climatic conditions, alternating with the movement of the sun between the tropics, was a conception that Eudoxus may well have had; but we wonder whether he would have described the regions known to him as "under the summer tropic" and the corresponding regions in the southern hemisphere as "under the winter tropic." [219] If so, he must be thought to have spoken with even less regard for accuracy than usual or to have been quite ignorant of the real location of the tropics. It seems more probable that we have here another datum derived from a work dating from Ptolemaic times,[220]

[217] Aristotle, *Liber de inundacione Nili* (fr. 248, Valentin Rose, *Aristotelis qui feruntur librorum fragmenta,* Leipzig, 1886, p. 195), reports the same opinion as from Nicagoras of Cyprus, not from Egyptian priests: "Nicagoras autem Ciprius ait ipsum [*sc.* Nilum] fluere amplius estate eo quod fontes habeat ex terra ad illam partem, in qua hyems est quando fuerit apud nos estas."

[218] Hippocrates, *De victu,* II, 38 (Émile Littré's edition, Paris, 1839-1861, Vol. 6, p. 532). For the related opinion of Aristotle, see *Meteorology,* 362 b 30-36.

[219] See Aëtius, III, 11, 4, where a similar expression is used in reference to Parmenides.

[220] The statement regarding Eudoxus has been repeatedly combined with one in Diodorus Siculus (I, 40, 1f.), where a similar view is attributed to "certain philosophers of Memphis." See P. Friedländer, "Die Anfänge der Erdkugelgeographie," in *Jahrbuch des Deutschen Archäologischen Instituts,* Vol. 29, 1914, pp. 117-118; Karl Reinhardt, *Parmenides und die Geschichte*
(*continued on next page*)

when we hear much of "the philosophers of Memphis," with whom Eudoxus is reported to have communed for a term varying in length from sixteen months to thirteen years. The whole may therefore be regarded as subject to justifiable suspicion.

The foregoing review of the available data regarding Eudoxus does not entitle us to draw a dogmatic conclusion. All that we may reasonably say is that Aristotle represents a stage in the development of scientific geography in essential agreement with that to which the data regarding Eudoxus point as having been attained in the second quarter of the fourth century. The fact that Eudoxus was concerned with the mathematical problems of spherics likewise suggests that we should look to that period as the one in which the implications of the sphericity of the earth, suggested some time earlier, are most likely to have been discerned and developed. If this view is correct, the threshold of scientific geography was reached in this period: the theory of the earth's sphericity had been essentially formulated, but the necessary data for the construction of an adequate map and for an estimate of the size of the earth were yet to seek.

Continuation of footnote 220.
der griechischen Philosophie, Bonn, 1916, pp. 147f.; W. Capelle, "Die Nilschwelle," in *Neue Jahrbücher für das klassische Altertum, Geschichte und deutsche Literatur,* Vol. 33, 1914, pp. 345-346. I have no doubt that the "Egyptian priests" and the "philosophers of Memphis" are alike fictions of the Ptolemaic age. There is nothing Egyptian about the conception.

CHAPTER VIII

DETERMINATIONS OF GEOGRAPHICAL POSITIONS IN THE PERIOD AFTER ARISTOTLE

The time was one that must have stimulated interest in these questions in an extraordinary measure; for not only had the conquests of Alexander opened up the east and brought to the knowledge of men of science a vast amount of significant information, but the rivalry of Alexander's successors—which was fortunately not confined to material interests—led to the employment of scholars and enquirers by the various courts in efforts to establish their renown. If we may judge by the immediate results in the field of geography the reports current in later times regarding Alexander's own contributions to science are at least greatly exaggerated [221]; for it is evident that those who attended him on his expedition to India were far from being up to date, depending in fact, if not by preference, on very early maps and accepting views that such men as Hecataeus had abandoned. We shall presently see, however, that data gathered in the time of Alexander's immediate successors left their mark on later maps.

Among the discoveries of the age one deserves to be singled out as of special importance, though for the time being it concerned astronomy alone. Eudemus, a pupil of Aristotle who devoted himself especially to the history of the mathematical sciences, placed the tropic at 24° from the equator.[222] Because of its convenience, Eratosthenes

---

[221] The personal contribution of Alexander to geography, if it may be called such, was in good part negative, correcting false notions current in his day. Among the most valuable data were those given by Nearchus, who used Alexander's itineraries and published the account of his own voyage from the Indus to the Persian Gulf.

[222] Theo Smyrnaeus, *On the Mathematical Knowledge Which is Needed to Read Plato*, Eduard Hiller's edit., Leipzig, 1878, p. 199. It is not clear whether Eudemus himself made the determination or merely reported it. See above, note 159.

104 THE FRAME OF THE ANCIENT GREEK MAPS

and Hipparchus continued to employ this figure, though the former had more closely approximated the actual distance. There is no reason to think that Eudemus made an application of this datum to the geographical zones. Everything indicates, rather, that, as far as observations for geographical purposes were undertaken in this period, they were for the most part vague or not expressed in terms that prove a knowledge of the mathematical principles involved.

The Geographical Position of India

How vague or inaccurate such observations could be is suggested by statements regarding India. Of the statement of Deïmachus that India lies between the autumnal equinox and the winter sunrise we have already spoken (see above, p. 48). Hipparchus, probably because he did not understand the point of view, which appears to have been that of the Ionian geographers, suggested [223] that Deïmachus meant not "winter" but "summer" sunrise, which would place India between the tropic of Cancer and the equator. Even if that conjecture were approved, the result would be true only of the southern half of the peninsula, and the inaccuracy would be glaring.[224] We learn also that Deïmachus contradicted the statement of Megasthenes that in southern India the shadow shifts and the Bear sets.[225] Megasthenes, who lived under Seleucus I and was employed by him as ambassador about 300 B. C., was quite right, though we could wish more definite data. His statement, however, was obviously based on a record of firsthand observations. Onesicritus, Alexander's chief pilot, who under the com-

[223] Strabo, II, 1, 19.
[224] Cape Comorin lies more than 8° north of the equator, and the tropic of Cancer lies just south of the mouth of the Indus.
[225] Strabo, II, 1, 19. Hipparchus knew that Megasthenes was right; see Strabo, II, 5, 35; II, 1, 20.

DETERMINATIONS OF POSITIONS

mand of Nearchus conducted the fleet from the Indus to the Persian Gulf, is reported [226] to have said that the sun at the summer solstice was directly over the Hypanis, one of the tributaries of the Indus. If he meant to be exact, he was in error; for the mouth of the Hypanis lies about 4° north of the tropic.[227] Unfortunately we do not learn to whom Eratosthenes and Hipparchus owed the information that placed the southern extremity of India and Ceylon on the parallel of Meroë and the Cinnamon Coast.[228] If, as has been surmised, it was a mere inference from climatic conditions [229] reported by traders, it constitutes further evidence of the lack of appreciation of accurate astronomical observations two centuries after the time of Aristotle; for trade with India and Ceylon was carried on at that time and became brisk a little later. We should suppose that if a few astronomical observations had been recorded, they could have been obtained from ship captains and would have enabled an astronomer to determine the latitude with approximate accuracy.

PHILO'S OBSERVATIONS ON THE RED SEA AND THE NILE

Astronomical observations, however, were obviously rare,[230] for even Eratosthenes and Hipparchus could cite but few. One, which must have been especially welcomed, was furnished by Philo, who seems [231] to have served as admiral under Ptolemy I (323-285 B. C.). His date is, indeed, not quite certain; but from the statement of Hipparchus, quoted by Strabo,[232] that Eratosthenes very nearly

---

[226] Pliny, *Naturalis historia*, II, 183. Contrast Herodotus, III, 104.
[227] Equally at fault is the datum of Pliny, *op. cit.*, II, 185.
[228] The data of Pliny, *op. cit.*, II, 182ff., are vague and anonymous.
[229] See Strabo, II, 1, 2, and Hugo Berger, *Die geographischen Fragmente des Eratosthenes*, Leipzig (Teubner), 1880, p. 191.
[230] See E. H. Bunbury, *A History of Ancient Geography*, London, 1879, Vol. 1, pp. 632f., 660f.; Karl Müllenhoff, *Deutsche Altertumskunde*, Berlin, 1887-1900, Vol. 1, p. 309.
[231] Pliny, *op. cit.*, XXXVII, 108.
[232] II, 1, 20. See Pliny, *op. cit.*, II, 183f., VI, 171.

agreed with the determination of Philo, we may conclude that his observation was made before the time of Eratosthenes.

In an account of a voyage to Ethiopia Philo stated that the sun is in the zenith at Meroë forty-five days before the summer solstice and he gave the ratio of the height of the gnomon to the length of its shadow at the equinoxes and the solstices. It is possible, even probable, that related data regarding Ptolemaïs Epitheras on the Red Sea were likewise derived from Philo, who as an admiral was more likely to have visited the seaports than the inland Meroë, about which he may have derived his information from traders or hunters. With the observations regarding Ptolemaïs Epitheras is coupled one for Berenice, also situated on the Red Sea and located on the tropic. Though sometimes attributed to Philo, this observation is probably of later date.[233] In any case, there were here provided verifiable data, which, though not absolutely accurate, admitted of being used by competent astronomers in determining latitude.

Observations of Pytheas in the North

While Philo's reports formed a basis for estimates or calculations, they related to southern lands that were comparatively little known. For the Mediterranean Basin, in which centered the chief interest of the Greeks, the earliest well attested observation available for the determination of latitude was made by Pytheas of Massalia, who was probably Philo's (perhaps older) contemporary. It is impossible to fix the date of Pytheas, the only evidence

---

[233] Hugo Berger (*Geschichte der wissenschaftlichen Erdkunde der Griechen*, 2nd edit., Leipzig, 1903, p. 413, note 3) attributes the observation for Berenice also to Philo; but Berenice was founded and named after his mother by Ptolemy Philadelphus (285-247 B. C.), probably about 250 B. C., for the earlier years of his reign were very much occupied with foreign wars. I incline, therefore, to attribute this observation to Timosthenes, the admiral of Philadelphus.

DETERMINATIONS OF POSITIONS 107

being that Dicaearchus, a pupil of Aristotle, is said [234] to have given no credit to the reports of his travels and observations in the north. Since we do not know how long Dicaearchus lived,[235] this statement affords us little help. We may, however, tentatively place Pytheas at about 300 B. C., or a little earlier. It is not necessary for our present purpose to estimate the value of his contribution to geography as a whole, about which opinions have differed and still differ, but there is no question of the importance of one observation with which he must be credited. He stated [236] that at his native city, Massalia (Marseilles), at the summer solstice the ratio of the length of the shadow to the height of the gnomon was as 120 to $41\frac{4}{5}$. Whether it was he or succeeding geographers, like Erastosthenes and Hipparchus, who deduced from this datum the latitude of the city, we do not know. The observation, however, not only is notably precise in form, but also approximates very closely to the truth. Though we may not say with assurance that Pytheas fully understood the mathematical relation between latitude and the meridian height of the sun or the lengths of the longest and shortest days, it is clear that he knew that such a relation existed and obviously saw the general bearing of the information he gave; for he reported data of the latter sort regarding localities in Gaul,[237] from which Hipparchus deduced their latitude. In like manner his statement that at Thule "the Arctic Circle and the northern tropic circle are the same" [238]

[234] Strabo, II, 4, 2.
[235] Possibly till 285 B. C.
[236] Strabo, II, 5, 41; see also I, 4, 4, where Hipparchus is said to have given the same ratio for Byzantium. The observation for Massalia is very nearly correct (Hipparchus made it a little more than 43° north latitude; actually it is 43° 21'. Hipparchus, however, made a considerable error in regard to Byzantium, which lies in 41° 1' north latitude).
[237] Strabo, II, 1, 17-18.
[238] Strabo, II, 5, 8. The term "Arctic Circle" is evidently used here in the sense of the "circle of perpetual apparition" (i. e. the circle in the heavens within which the stars never touch the horizon—consequently a circle that
(continued on next page)

shows that he understood the relation between the latitude of a location on the earth's surface and the lengths of the longest days and nights. That the nights were long in the far north had, of course, long been known; but the precise mathematical formulas for converting vague data of this sort into terms of latitude were evidently still to be learned.

Whether Pytheas in making his observations was moved by a desire to determine the size or circumference of the earth there is no way of telling. Speculation on the subject was certainly natural enough in his time; but the want of information regarding his purpose and especially of any indication that he gave definite estimates of distances from point to point does not favor the assumption that such was his motive. One would rather conclude that he was chiefly concerned with general geographical exploration and with evidence confirming the sphericity of the earth. In short, he would appear to have been an observer rather than a calculator, as has been said of Eudoxus, whom some scholars suppose to have been his teacher. If this conclusion is sound, it does not in the least derogate from

Continuation of footnote 238.
varies with the latitude of the observer). Nevertheless, the assumed coincidence of this circle with the summer tropic presupposes that Thule was on or near the fixed geographical Arctic Circle. From any point on the latter, looking north, one could see the sign of the Crab and, looking south, the sign of the Goat. This interpretation of the statement in Strabo would appear to be confirmed by the following passages from Cleomedes. In his *De motu circulari corporum caelestium* (I, 7; Hermann Ziegler's edit., Leipzig (Teubner), 1891, p. 68, lines 19-26) Cleomedes says: " It is said that the so-called island of Thule, where Pytheas, the Massiliote philosopher, is reported to have been, the whole summer tropic is above the horizon, it becoming for the inhabitants the Arctic Circle. There, when the sun is in the Crab, the day attains the length of a month, if all parts of the sign of the Crab are always visible there, or rather so long as the sun is among the stars that are always visible there." Elsewhere we read: " For the inhabitants of Thule the summer tropic coincides with the Arctic Circle; for those who live farther [north] the Arctic Circle goes proportionally beyond the summer tropic toward the [celestial] equator; at the pole itself the equator has three functions, being the Arctic Circle, in that it embraces those stars that are ever visible, none there rising or setting; it serves also as the horizon, dividing the hemisphere of the cosmos above the earth from that below the earth; and the equinoctial circle also, because it there divides day and night into equal parts " (*ibid.,* Ziegler's edit., p. 70, lines 10-22).

DETERMINATIONS OF POSITIONS

the significance and importance of his work but rather indicates a sober appreciation on his part of the problems immediately pressing in the field of geography. He was, to be sure, accused of drawing the long bow in his accounts of the north, but it is difficult to judge whether he did so or not, since the evidence is quite inadequate both as to what he actually said and as to the actual facts in the lands he visited. If he did, he was using a time-honored privilege of the traveler.

The data furnished by Philo and Pytheas were of fundamental importance, if they were appreciated in their time. Especially important would have been the location of Berenice on the tropic, if we could only be quite sure that it was actually made by Philo about 300 B. C.[239] We cannot, however, help questioning this inference drawn from the statement of Pliny, because there is no evidence that anyone made use of this datum in estimating the circumference of the earth. The circumstance is apparently significant that one hears nothing of the location of Berenice on the tropic until long after Syene, situated on the same parallel, had come to play its leading role. If we accept the view that Philo reported the fact, we must apparently conclude either that its importance was not fully appreciated or that another point of reference, Syene, was known, one of equal value and practically more accessible.[240]

[239] See above, note 233.
[240] It is possible that the location of Syene (almost) on the tropic was recognized after that of Berenice. The latter would be noted by seamen who were by that time familiar with the use of the gnomon, while Syene would naturally be visited by soldiers and merchants, less concerned with questions of latitude. Moreover, the overland connections of Berenice were not with Syene, but with Coptus.

CHAPTER IX

DICAEARCHUS

It has become the fashion to attribute to Dicaearchus the first utilization of Syene in estimating the circumference of the earth. Dicaearchus, a native of Messana in Sicily, was a pupil of Aristotle and a fellow pupil of Aristoxenus. He must, therefore, have been in his prime about 300 B. C. or a little earlier. As we have seen, he is said to have declined to credit the reports of Pytheas. If we knew on what grounds he rejected them, we should be in a better position to set a value on his own geographical work. That he was a man of parts there can be no question; and for our study he is important because he occupied himself seriously with geography.

Regarding his views and achievements in this field we actually know surprisingly little, especially when we consider that he was numbered among the notable geographers.[241] That he prepared a general map may be accepted as probable, if not quite certain; but we are ignorant of how it may have looked.[242] He of course accepted the view that the earth is spherical[243] and doubtless knew and accepted the determination of the obliquity of the ecliptic at 24° reported by his fellow disciple, Eudemus. How this may have affected his map we cannot say; we know, however, that he gave the ratio of the length of the *oikumene* to its breadth as 3 to 2, thus agreeing with Democritus,[244] although we do not know how he arrived at this ratio. At

---

[241] Strabo, I, 1, 1; II, 4, 1; Agathemerus, I, 2 (Carl Müller, *Geographi graeci minores,* Paris, 1882, vol. 2, p. 471).
[242] Except for Greece; see Cicero, *Ad Atticum,* VI, 2, 3.
[243] Pliny, *Naturalis historia,* II, 162.
[244] Agathemerus, I, 1.

first sight it might seem that he was harking back to Ionian ideas, which would not be surprising in view of the interest he displayed in ethnology and kindred subjects cultivated by the Ionians. His treatise on the *Life of Hellas* was clearly inspired by Ionian models. Another curious datum suggests the same influences; we are told that he explained the floodwaters of the Nile as coming from the ocean. We may doubt that he made this statement, but if we accept it as his, it is surely more natural to believe that he was harking back to Hecataeus of Miletus and Euthymenes of Massalia than to suppose that he was thinking of torrential rains brought from the ocean by the southwest monsoon, as Berger suggests.[245] However, it is hazardous to draw any positive conclusion from the scanty evidence; for Dicaearchus must have had at his disposal considerable information regarding the extent of the *oikumene* gathered in the course of Alexander's campaign, even if he did not use the official itineraries published by Nearchus. What we learn about his estimates of distances in the near-by Mediterranean basin[246] does not redound especially to his credit. We know also that he estimated the altitude of certain mountains in Greece[247]: the inaccuracy of his results[248] hardly justifies us in saying that he measured them. It is probable that he was concerned to determine how these elevations affected the rotundity of the earth, but this obviously does not imply any particular computation of its circumference.

The Median Axis of the "Oikumene"

Regarding the structure of his map we have only one certain indication. Agathemerus says[249]: "Dicaearchus does

---

[245] *Geschichte der wissenschaftlichen Erdkunde der Griechen*, 2nd edit., Leipzig, 1903, p. 377, note 1.
[246] Strabo, II, 4, 2.
[247] Pliny, *op. cit.*, II, 162.
[248] See E. H. Bunbury, *A History of Ancient Geography*, London, 1879, Vol. 1, pp. 617f.
[249] I, 5 (Müller, *op. cit.*, p. 472).

not bound the earth by waters, but by a straight line drawn from the Pillars through Sardinia, Sicily, Peloponnesus, Ionia, Caria, Lycia, Pamphylia, Cilicia, and the Taurus, one after the other, up to the Imaus Mountains. Of these regions he calls the one the northern, the other the southern." The sentence as a whole is rather stupid. Dicaearchus referred, no doubt, to the *oikumene,* and we are probably to understand that he did not divide it into continents, whose limits were marked by rivers, straits, an isthmus, etc., but separated the northern from the southern half of the land mass (no doubt regarded as surrounded by water) by an imaginary line running from west to east. Making due allowance for vagueness of expression, we have a fairly satisfactory line roughly parallel to the equator.

This line, we know, was adopted by Eratosthenes as the main longitudinal axis of his chart.[250] As was pointed out above, it practically coincided with the equator of the Ionian map between the Pillars and Asia Minor. From thence the text of Herodotus based on the " Persian map " indicates a line, marking the boundary between the tract (ἀκτή) of Asia Minor and that of Syria, in the same direction as far as the eastern limits of Persia. Herodotus does not say that it follows the trend of the Taurus, as Dicaearchus and Eratosthenes represented it, but that was the natural boundary at least at its western end. There were apparently differences in the dimensions given to Asia Minor,[251] but the ground plan of the map in this region remained essentially unchanged.

[250] Strabo, II, 1, 1.

[251] Herodotus (II, 34) thought that a well-girt traveler could traverse the distance from the Cilician Gates to Sinope in five days, which would make the distance little more than 100 miles. According to Strabo (II, 1, 3; XI, 1, 3), Eratosthenes reckoned the breadth of the Taurus (or mountains of Asia Minor) at 300 miles. Apparently Dicaearchus agreed more nearly with the estimate of Herodotus.

DICAEARCHUS

It is at the extreme east that important differences appear between the Ionian maps and the maps of Dicaearchus and of Eratosthenes. The historians of Alexander, who depended on the maps of the Ionians, regularly speak of the Hindu Kush and the Himalayas as the Caucasus,[252] while Dicaearchus and Eratosthenes, except when they are quoting the Macedonians, call them the Taurus. Eratosthenes, moreover, in a passage already cited,[253] says that the old map showed India too far north. On what evidence Dicaearchus made the change, placing the northern boundary of India on his equatorial axis we do not know; but we may conjecture that it was chiefly on climatic grounds, such as led Eratosthenes to place the southern limit of India on the parallel of Meroë. How inexact these data were, we have already seen.

From the above it is clear why in our account of the disk earth [254] we raised the question whether Deïmachus may not actually have been reproducing the map, if not the language, of his contemporary, Dicaearchus, when he said that India lies between the equinoctial and the winter sunrise.

Did Dicaearchus Estimate the Circumference of the Earth?

The details thus far presented comprise the sum of what may fairly be regarded as known about the geographical work of Dicaearchus. We are naturally not satisfied with this meager fare when dealing with a disciple

---

[252] Conversely, Avienus, 250, gives the name Taurus to the mountains beginning at the Cimmerian Bosporus. This is doubtless significant and suggests for this portion of his account a source dating at the earliest from the end of the fourth century, because it was not until after Alexander's eastern campaign that the series of mountain chains crossing Asia from west to east came to be called Taurus instead of Caucasus, as in earlier times.
[253] See above, note 107.
[254] See above, p. 49.

of Aristotle who was held in great esteem not only for character but for learning and for contributions to many subjects, including geography. It was inevitable, therefore, that conjecture and assertion should supplement the record. While we may ignore most of these unverifiable assumptions, one of them has won so many supporters that we are bound to consider it. Berger,[255] whose history of the scientific geography of the Greeks justly enjoys an honorable reputation, has persuaded himself and others that Dicaearchus was the author of an estimate of the circumference of the earth.

His conjecture rests essentially on two data. Cleomedes in his *De motu circulari*[256] says: " The inhabitants of Lysimachia have the head of the Dragon overhead, while over the region of Syene stands the Crab. Now, as the sundials show, the arc measured from the Dragon to the Crab is one fifteenth of the meridian that passes through Lysimachia and Syene; but one fifteenth of the whole circle is approximately one fifth of its diameter. If, then, we assume that the [surface of the] earth is even and draw perpendiculars to it from the extreme parts of the periphery extending from the Dragon to the Crab, they will touch the diameter of the meridian passing through Lysimachia and Syene. Now the distance between the perpendiculars is 20,000 stades; and since this is one fifth of the whole diameter, the diameter of the meridian is 100,000 stades, and the circumference will be 300,000 stades." We have, then, a computation of the circumference of the earth at 300,000 stades, based on the distance between Lysimachia and Syene, which is given as 20,000 stades, and on the observations that the Crab and the head of the Dragon are vertical

---

[255] *Die geographischen Fragmente des Eratosthenes,* Leipzig (Teubner), 1880, pp. 107, 173-178; *Geschichte der wissenschaftlichen Erdkunde der Griechen,* 2nd edit., Leipzig, 1903, pp. 370-373, and *passim.*

[256] I, 8 (Hermann Ziegler's edit., Leipzig (Teubner), 1891, p. 78).

DICAEARCHUS

respectively over Syene and Lysimachia. There is no intimation to whom we should credit this estimate.

The second datum is a statement of Archimedes in his *Sand Reckoner* [257]: "In the first place, I assume that the circuit of the earth is 3,000,000 stades, and no more, though there are, as you know, those who have tried to prove that it is 200,000; but, going beyond that and taking the size of the earth as tenfold that which was supposed by those who formerly have expressed an opinion, I assume it to be 3,000,000 stades, and no more." The purpose of Archimedes is evident. Wishing to expound and show the possibilities of his new system of numerical notation, he assumed an incredibly large volume of the earth, the sands of which his method would still suffice to number.

Berger points out that Archimedes refers to the estimate of 300,000 stades (the figure that is also given by Cleomedes) as having been "formerly" made, and he couples with this the fact that Lysimachia was founded by Lysimachus in 309 B. C., when Dicaearchus was in his prime. He assumes that the estimate could not have dated from before that time and concludes that either it must have been made by Dicaearchus or Dicaearchus had a part in the undertaking. This is the essence of his argument, though he seeks to eke out these data with scraps of information regarding other thinkers, such as the determination of the inclination of the ecliptic reported by Eudemus as 24° (=1/15 of the great circle).[258] Eudemus, to be sure, was a contemporary and fellow pupil of Dicaearchus, but we do not see what bearing this has on the question who made the estimate of the circumference of the earth.

---

[257] J. L. Heiberg, *Archimedis opera omnia*, 2nd edit., Leipzig, 1910-1915, Vol. 2, p. 220.

[258] Since this figure was accepted not only by Eratosthenes and Hipparchus, but throughout later antiquity, it cannot serve to date the estimate of the earth's circumference. The contention that the estimate, because of its incorrectness and the inaccuracy of the data used, must have been made before Eratosthenes is not cogent but may be allowed to have some weight.

At first blush Berger's conjecture is attractive. If it were attested by our sources, we should not hesitate to attribute this computation to Dicaearchus; but it is far from being established and should be approved, if at all, only as a possibility. The difficulty is only partly due to the fact that we have so little positive knowledge of Dicaearchus. As far as Berger's conjecture is concerned, its chief reliance is on the statement of Archimedes, which refers to the past the estimate of 300,000 stades for the circumference of the earth. There is no way of telling what date Archimedes had in mind because we do not know definitely when he wrote the *Sand Reckoner*.[259] This work was dedicated to Gelon (son of Hieron II), who died 216-215 B. C., about four years before the death of Archimedes. Gelon was probably approximately contemporary with Eratosthenes, who was born about 276 B. C., and we may assume that Archimedes addressed him in his maturity. We should then date the *Sand Reckoner,* say, about 230-216 B. C. On this assumption Dicaearchus would date three quarters of a century earlier; and the term " formerly " might certainly refer to a much shorter period. Besides, Archimedes seems to have been little concerned about the history of the problem: he mentions also an estimate of 200,000 stades as known, but he does not indicate whose it was. If he referred to Eratosthenes, as has been conjectured, he was not at all exact, for the precise figure of Eratosthenes was 252,000 stades. He implies, moreover, that the estimate of 300,000 stades was the largest that had been made, though we know that Aristotle reported one of 400,000. The figure 200,000 also might obviously with equal right have been referred to as a " former " estimate. From these considerations we may infer that Archimedes was chiefly concerned to take

[259] Sir Thomas Heath (*Aristarchus of Samos* . . . , Oxford, 1913, p. 337) accepts Berger's conjecture and would fix no date for the *Sand Reckoner* except "before 216 B. C."

a figure that might be regarded as the maximum current at the time, and this figure he multiplied by ten in order to give no opportunity for quibbling. On this score, therefore, there appears to be no good reason for thinking especially of Dicaearchus in this connection.

Berger makes a point of the date of the founding of Lysimachia, 309 B. C., which fell about the prime of Dicaearchus. There is no reason to suppose that this event, occurring at a time when many cities were being founded, claimed especial attention. In particular, the occasion offers no apparent ground for considering the relation of the city to Egypt. During the following years, to be sure, Lysimachia, because it commanded the Hellespont, played an important role in the wars between Ptolemy and the other generals of Alexander, Demetrius and Antigonus, and experienced vicissitudes more likely to lead to an estimate of the distance between it and points in Egypt. It is idle, however, to discuss these possibilities, because we do not know that Archimedes referred to this particular calculation. What reason have we to suppose that in the third century before Christ there were no more than one or two attempts to solve a problem that must have engaged the thought of every mathematician and astronomer?

The calculation reported by Cleomedes is assumed to be older than that of Eratosthenes, because it is less accurate than his. Though this is probable, it is not certain. The calculation of the earth's circumference is based upon the supposition that the head of the Dragon is vertical over Lysimachia and [the center of?] the Crab over Syene and upon an estimate, very rough indeed, of the distance between these cities. Cleomedes knew that Syene lies on the tropic of Cancer. The problem, therefore, resolves itself into the question when was this discovery made. The fact was known to Eratosthenes, who based his calculation on it. Was it known a century earlier? In view of the almost feverish activity of science in Egypt during the third century it seems to me hardly credible that a century should have elapsed without an attempt to arrive at an estimate of the circumference of the earth based on more exact data. I think it most probable that the calculation based on the observations at Lysimachia and Syene dates from the time of Eratosthenes and that it provoked him to improve upon it.[260] The observation at Lysimachia was very inexact, since the latitude there is 40°35' and the declination of γ Draconis is 53°.[261] This may be regarded as evidence for an early date; but we must note that the latitude of Lysimachia was determined by the zenith position of a group of stars, a method that, as employed at least in the earlier time, gave far less accurate results than the use of the gnomon in measuring the meridian height of the sun, the method that was doubtless employed in fixing the latitude of Syene. Although the language of Cleomedes suggests that the sundial was employed in determining the arc between the Dragon and the Crab, it is evident that the position of the Dragon could not be so measured. For the Crab the gnomon, or its equivalent, would be the natural instrument to use. The fact that two methods were presumably employed suggests to me that the data may not have been due to the same person. Hence it is quite possible that an old datum regarding Lysimachia, together with an approximate estimate of the distance between the two cities, was combined

---

[260] Eratosthenes located Lysimachia on the same parallel with Sinope (42° 1'). See Strabo, II, 5, 40.

[261] My colleague, Professor Frederick Slocum, has kindly checked the data for me: "If we take for the head of the Dragon the star in the temple (γ Draconis), which is the brightest star in the head, we find its declination to be 53°. This would pass 12½° to the north of the zenith of Lysimachia. Other stars in the head would pass still farther to the north." Hipparchus (*In Arati et Eudoxi Phaenomena commentariorum*, I, iv, 8; Carl Manitius' edit., Leipzig, 1894, pp. 34, 10ff.) says: "The star in the tip of the Dragon's mouth is 34¾° from the pole, its southern eye is 35°, its southern temple 37°." Professor Slocum assures me that this is approximately correct for the time of Hipparchus.

with the datum regarding Syene. In that case there is nothing in the procedure to date the calculation, except that it was presumably earlier than that of Eratosthenes. There is no evidence regarding the date of the discovery that Syene lies approximately on the tropic; we know only that this was recognized by Eratosthenes in his calculation of the circumference of the earth. In his early poem, entitled *Hermes,* Eratosthenes himself had betrayed no knowledge of this fact.[262] It seems significant, moreover, that the distance between Meroë and Syene was always estimated on the basis of the length of a degree of latitude as determined by Eratosthenes. If, as seems most probable, the location of Berenice on the tropic was recognized in the time of Ptolemy Philadelphus (285-247 B. C.), we may assume that the geographical position of Syene was a later discovery. Once the latitude of Syene was determined, the fact must have seemed especially important; for Syene lies on the Nile, whose course thence to Alexandria runs approximately north, making it feasible to disregard possible differences of longitude. Moreover, the distance overland between Syene and Alexandria could be given more accurately than between Syene and Lysimachia, because it was not necessary to depend on vague estimates of days' sailings. It is not unreasonable, therefore, to suppose that Eratosthenes, stimulated by the attempt to calculate the circumference of the earth on the basis of the distance between these cities, sought to improve upon it by taking the distance between Syene and Alexandria as the basis of his estimate. The greatest improvement in his method,

---

[262] See Karl Müllenhoff, *Deutsche Altertumskunde,* Berlin, 1887-1900, Vol. 1, p. 244; Hugo Berger, *Geschichte der Wissenschaftlichen Erdkunde der Griechen,* 2nd edit., Leipzig, 1903, p. 394. A. Thalamas (*La géographie d'Ératosthène,* Paris, 1921, pp. 174-175) maintains that the *Hermes* presents the same picture of the earth as the *Geography.* This is undoubtedly true, so far as it concerns the astronomical and mathematical theory; but the poem does not indicate the geographical location of the zones.

however, consisted in the use of the gnomon in determining the latitude of both cities at the ends of his base line.

These considerations justify us in withholding assent to Berger's conjecture, though it has been approved and accepted by eminent scholars. It may be admitted as a possibility, but hardly as more. Consequently the contribution of Dicaearchus to geographical science remains for us extremely vague. Except in the Mediterranean basin and along his longitudinal axis the available information affords us no idea of his map, and even in these regions there is nothing to suggest a notable improvement on the work of his predecessors.[263] Such advances as he made seem to be largely confined to the east and may with probability be explained as due to reports of companions of Alexander.

If we are to offer a conjecture regarding the author of the estimate of the earth's circumference reported by Archimedes and Cleomedes, it would seem more reasonable to think, as Bunbury did, of Aristarchus of Samos,[264] for Aristarchus is several times mentioned in the *Sand Reckoner* and there is every reason to think that he had formed some definite conception of the size of the earth, not only because he was concerned in determining the relative sizes of the sun and moon, but also because he regarded the size of the earth as negligible in comparison with the universe. If he proposed the figure 300,000, we fancy that he did so somewhat in the spirit of Archimedes—that is to say, as a maximum, without pretending that it was accurate. For such a rough estimate the data regarding

---

[263] See E. H. Bunbury, *A History of Ancient Geography*, London, 1879, Vol., 1, p. 661, for the error of Dicaearchus, partly corrected by Eratosthenes, regarding the Strait of Messina.

[264] See Bunbury, *op. cit.*, Vol. 1, pp. 620-621. According to Ptolemy (*Syntaxis mathematica*, III, 1; J. L. Heiberg's edition of the works of Ptolemy, Leipzig (Teubner), 1898-1903, Vol. 1, pp. 206-207), Aristarchus made an observation of the time of the summer solstice in 281-280 B. C., the year in which the Thracians destroyed Lysimachia.

Lysimachia and Syene and the distance between them would suffice. We have no reason, however, for thinking of Aristarchus as specially interested in the details of geography; we should rather regard him as following in the footsteps of Eudoxus, applying the advanced knowledge of mathematics to the problems his predecessor had broached and finding a better solution of the apparent movements of the celestial bodies.

Our conclusion, then, is that Dicaearchus remains for us largely unknown, while Aristarchus and Archimedes, from whom we might have expected much in the furtherance of scientific geography, were primarily interested in other things.

CHAPTER X

ERATOSTHENES

With Eratosthenes we enter into a new world. An encyclopedist in the scope of his interests and studies, he devoted himself with especial energy to geography, sketching the history of the earlier essays in this field and seeking above all to perfect the map of the earth. Having at his disposal a wealth of materials [265] literary and cartographic, such as none of his predecessors had possessed or at least had utilized, and being in addition a competent critic of sources, he was as it were by nature designated as the historian of the science. The loss of the first book of his geographical treatise is no doubt greatly to be regretted; but we may infer from the references to it, chiefly by Strabo, that the historical review was extremely sketchy, being in good part devoted to a criticism of the exaggerated notions of certain Stoics regarding the geographical knowledge of Homer. If he mentioned details about his predecessors, aside from lists of geographers and cartographers and an occasional judgment concerning the genuineness of the works attributed to them, we have no indication of them in our sources. What he gave was apparently dictated by the questions that were topics of current discussion.

Eratosthenes' Map

Regarding his contributions to geography it is hardly necessary to speak in detail, for the subject has often been discussed with ample fullness and accuracy. We are concerned only with his influence on the map. That it was

[265] Strabo, II, 1, 5.

decisive and that his map formed the basis of all later maps of the ancients is acknowledged by all. Such changes as were made by his successors affected details only. For the improvement of the map he could avail himself not only of the increase of knowledge resulting from the campaigns of Alexander but also of the explorations undertaken by Pytheas and the reports of ambassadors and admirals in the service of the kings of Syria and Egypt. That he was generally judicious in the selection and use of his authorities is certain, despite the criticisms of Hipparchus and Strabo. He could not, however, escape the limitations of the knowledge of his day, and in the absence of adequate information he was compelled to choose the accounts that seemed to him the most probable. In doing so he was of course influenced by preconceptions, the origin and nature of which are not always obvious.

In drawing the frame of his map, however, he was on relatively firm ground. He was convinced that the earth was spherical, and he was at pains to present proofs for his thesis. Strabo thought he did so at needless length; but we infer that Eratosthenes would not have agreed with him and that in his day there was justification for arguing the point. At all events, he did so, for his conception of the earth depended on this assumption and upon an estimate he had made of the size of the globe.

The chief merit of Eratosthenes is that he not only accepted the view that the earth is a sphere and presented the arguments that could then be urged to prove it but also, utilizing such data as existed or could be obtained by his own observation, made a calculation of the earth's circumference that closely approximates the truth. No doubt he dwelt on this subject in various parts of his general geographical treatise, and we now know that he devoted a special work to the question. Hero of Alexandria, who flourished about the middle of the third century of our era, in his *Dioptrics*[266] cites this work, in which Eratosthenes presumably first announced his result and explained his method. Galen thus summarizes its contents [267]: "In it Eratosthenes treated of the size of the equator, the distance from it of the tropic and polar circles, the size of the polar zones, the sizes and distances of the sun and moon, their total and partial eclipses, the changes in the length of days depending on season and latitude—in short, all that we include in astronomical and mathematical geography." We see at once that Eratosthenes dealt with his subject in a most comprehensive way, approaching it from the point of view of the astronomer. It is natural that Hero should mention the treatise of Eratosthenes in his own work on the *dioptra,* a sort of sextant or theodolite with which one could measure angles, for which purpose Eratosthenes had used the sundial.

It is hardly necessary to explain in detail the procedure of Eratosthenes.[268] As we have said, he accepted the datum that Syene lies on the tropic of Cancer, which means that at the summer solstice the sun at midday is directly vertical over that point. By measuring the length of the shadow cast by the sun at Alexandria on that day, he computed the angular distance between the two points of reference, and, assuming that these points were on the same meridian and that the distance between the two cities as given by official surveyors was correct, he arrived at the figure 252,000 stades for the circumference of the earth. The result was too large by about $13\frac{3}{4}$ per cent,

[266] Chapter 35 (in Hermann Schöne's edition of the works of Hero of Alexandria, Leipzig, 1899-1914, Vol. 3). For a description of the *dioptra,* see M. Cantor, *Vorlesungen über Geschichte der Mathematik,* 4th edit., Leipzig, 1922, Vol. 1, pp. 382-383.
[267] *Institutio logica,* 12 (Karl Kalbfleisch's edition, Leipzig, 1896).
[268] Many textbooks on astronomy give the necessary information; *e. g.,* H. N. Russell, R. S. Dugan, and J. Q. Stewart, *Astronomy,* Vol. 1, Boston, 1926, p. 113. See also A. Thalamas, *La géographie d'Eratosthène,* Versailles, 1921, pp. 128-164.

but it was the closest approximation attained in antiquity, and it was accepted by Hipparchus. Various factors contributed to affect the accuracy of the calculation: Syene is not actually on the tropic but approximately 5' north of it; Syene and Alexandria are not on the same meridian, Alexandria lying 3°4' farther west; the official figure for the distance between the cities was too large, presumably because the surveyors who " paced " it off followed more or less the winding course of the Nile. Another factor of no small importance is the inaccuracy of the instrument used by Eratosthenes. The shadow of the gnomon could not be precisely measured because the rays coming from different parts of the sun's face do not follow parallel lines and therefore the shadow is blurred around the edges, just as in an eclipse there is a penumbra as well as an umbra. Furthermore, Eratosthenes himself recognized that the shadow would not perceptibly change within 300 stades, though the change would be marked at 400. That is to say, Eratosthenes allowed for an error of half a degree.

We know that Eratosthenes made a map,[269] and we are sure that it was the first map in which definite cognizance was taken of the sphericity of the earth. Just how did his epoch-making work affect the picture of the earth as it had been depicted by his predecessors? It is hazardous to attempt a reconstruction of his map in detail, but we have some precious information regarding it, which suffices to reveal his procedure. We know that he placed the whole *oikumene* north of the equator, to which we can with reasonable assurance trace at least seven parallels. One of these passed through Meroë and the southern end of India; the next ran through Syene and the Persian Gulf;

[269] There have been many attempts to reconstruct the map of Eratosthenes. One that is on the whole satisfactory may be conveniently consulted in the maps (No. 1) appended to the edition of Strabo by Carl Müller, Paris, 1880. This and other modern reconstructions, however, cannot be regarded as accurate in points of detail. See also Thalamas, *op. cit.*, pp. 187-251.

the third through Alexandria; the fourth from the Pillars of Herakles along the axis of the Mediterranean to Rhodes, thence along the "Taurus" to the northern boundary of India. This, as we have seen, was the main axis of the Ionian maps in the west and of some Ionian maps—those based on the "Persian map"—for a part of the east, at least as far as the eastern limit of Persia. It was the *diaphragma* of Dicaearchus, and it was likewise the main longitudinal axis of Eratosthenes' map. Farther north there was a fifth parallel, beginning in the west at the Pyrenees, skirting the northern end of the Gulf of Genoa, and passing through Byzantium and the mouth of the Phasis. A sixth parallel ran from a point just south of Britain through the mouth of the Maeotis and the upper part of the Caspian. Finally, Eratosthenes accepted the statement of Pytheas that "Thule" lay on the Arctic Circle. The distances between these parallels were given in stades, reckoning 700 stades to a degree of latitude. If we inquire how nearly the resulting locations of the principal features of the map along these parallels correspond to their positions as at present determined, we find that in the west between the parallels of Syene and Massalia the error is on the whole not great. Eastward of the Mediterranean, however, we find marked discrepancies, which are no doubt due to the fact that astronomical observations in that region were few and inaccurate, reliance being placed chiefly on reports of climatic conditions. In this respect, therefore, the picture of the *oikumene* presented by Eratosthenes, though showing appreciable improvement in many details, did not differ radically from that given by the Ionian map of the fifth century, as we have reconstructed its outline.

We have seen that the Ionian map also was not wanting in coördinates by which points could be located. Its chief parallels continued to appear on the maps, though they were found to be in need of some revision. No considerable change was made in regard to the Ionian equator, which followed the axis of the Mediterranean; and, as we have seen, Eratosthenes retained this line as the main longitudinal axis of the *oikumene*. The old lines connecting the sunrises and sunsets at the solstices were likewise retained, although between each of them and the Ionian equator Eratosthenes introduced a new parallel—one passing through Massalia and the other through Alexandria. He drew the Ionian summer tropic, as of old, along the northern end of the Euxine, through the mouth of the Maeotis and the beginning of the Caucasus in that region, though he made two important changes: the true direction of the Caucasus range, from northwest to southeast, was recognized; and, at the western end, the Pyrenees, now evidently located from the side of Massalia, were placed on a parallel with that city and thus shown much farther south. He still supposed the Ister to rise in the far west and to approximate the old Ionian summer tropic, and the Adriatic also approached that line. The winter tropic of the Ionians, which, as we have seen, ran through or close to Syene, reappeared on the map of Eratosthenes as one of his chief parallels, but, like the corresponding northern line, its significance was of course entirely changed. Philo's reports about Meroë had prepared the way, and by the time of Eratosthenes the location of Berenice and Syene on the tropic of Cancer was known. It is singular that there does not appear anywhere in our sources a reference to the astonishing discovery that the summer tropic actually lies almost exactly where the Ionians had placed their winter tropic. Yet it must have caused a sensation among geographers!

The Ionian maps, as we have seen, likewise indicate certain meridians, though few are definitely known. The one most clearly traceable runs from the Pelusiac mouth of the Nile by way of the Cilician Gates and Sinope to the mouth of the Ister. This line actually departs no more from the true meridian than that which Eratosthenes drew through Alexandria and Rhodes. The map of Eratosthenes shows other meridians, the most accurate being the one on which he located Rome and Carthage. In general the evidence is not sufficient to enable us to compare his locations in longitude with those of the Ionians. We know, however, that he conceived the length of the *oikumene* as greater in proportion to its breadth than did his predecessors, saying that it was more than twice as long as broad. For longitude Eratosthenes had still to depend on dead reckoning, for no astronomical observations were available. There is, indeed, no evidence that he had even conceived of a method of accurately determining longitude. It was Hipparchus who suggested the timing of eclipses for this purpose, but even he had no available data.

CONCLUSION

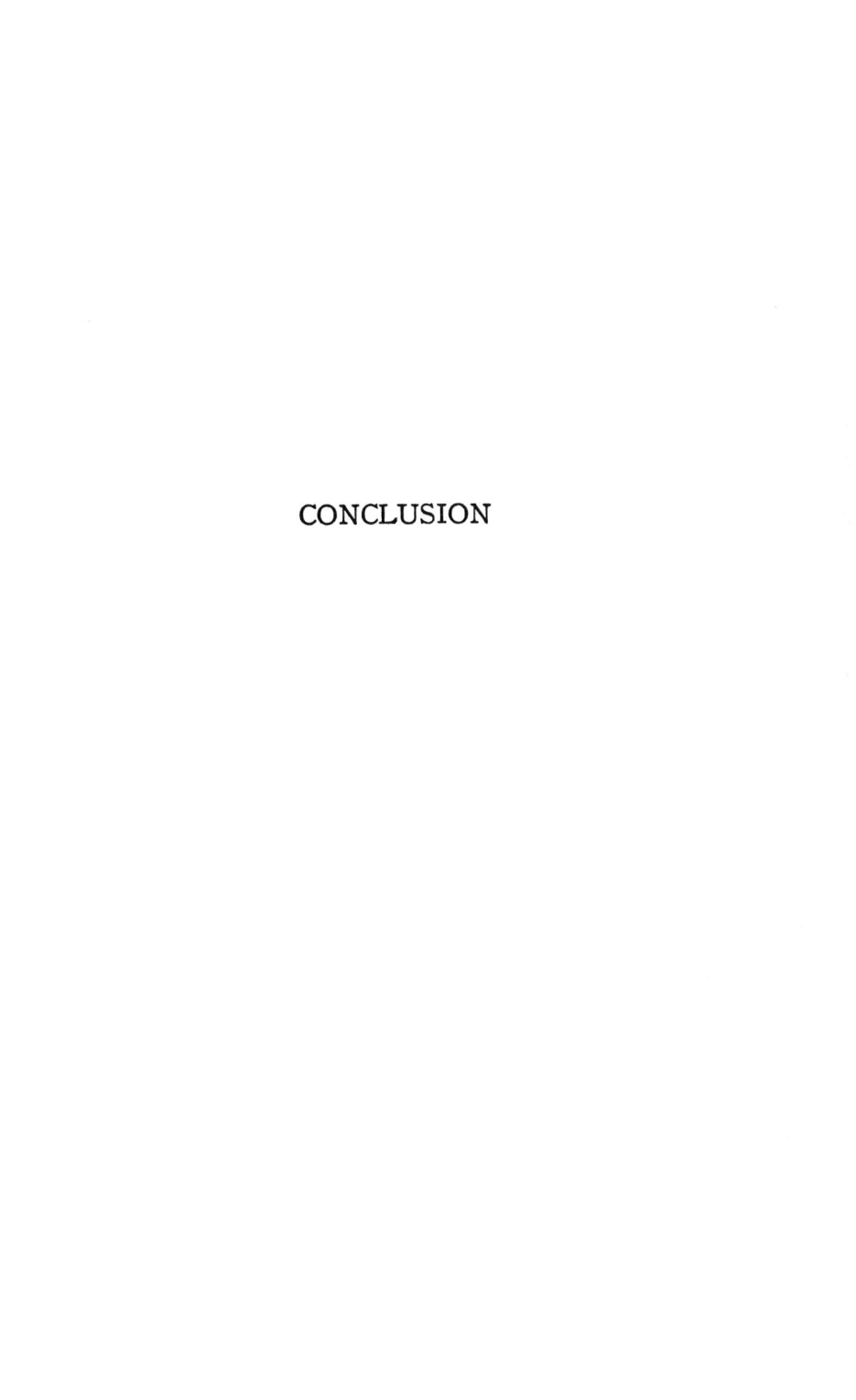

CHAPTER XI

THE ENDURING FRAME OF THE IONIAN MAPS

We may close our survey with Eratosthenes. Hipparchus, Marinus, and Ptolemy made important contributions in detail, but not all of their changes were by way of improvement. Eratosthenes was the first to attempt a map of the world based on astronomical observations and on the recognition of the sphericity of the earth, and in essentials the map remained as he left it. As we have already stated, Eratosthenes also sketched the history of cartography. How fully he treated the subject we do not know; but we surmise that he must have given relatively few details. Herodotus in the latter half of the fifth century mockingly referred to "many who had drawn maps of the world," but mentioned none by name. No doubt Eratosthenes himself found in the collections of the Alexandrian library a considerable number of maps. Who had drawn them? It is unlikely that their authors could always be identified. Since Herodotus nowhere indicates notable differences between the maps he knew and since Aristotle speaks in almost identical terms of those that were being produced in his time, we gather that no great changes had been introduced down to the latter part of the fourth century. Maps appear to have been produced in considerable numbers to satisfy the demand, and it is not likely that each was supervised by a geographer. Under such circumstances certain standard works would naturally be reproduced over and over again with only minor changes, as is the case today in the multiplying textbooks for school and college use.

Strabo says: "And, as everyone knows, the successors of Homer in geography were also notable men and familiar with philosophy. Eratosthenes declares that the first two successors of Homer were Anaximander, a pupil and fellow-citizen of Thales, and Hecataeus of Miletus; that Anaximander was the first to publish a geographical map, and that Hecataeus left behind him a work on geography, a work believed to be his by reason of its similarity to his other writings." [270] It is now agreed that this statement is based on Eratosthenes' sketch of the history of geography, from which is also derived a fuller account in the *Outline of Geography* of Agathemerus [271]: "Anaximander the Milesian, a pupil of Thales, first ventured to depict the *oikumene* on a tablet. After him Hecataeus of Miletus, who had traveled far and wide, treated the same subject so accurately that his work was regarded with wonder. Hellanicus of Lesbos, indeed, issued his account without a graphic representation. Then Damastes of Sige made a map, for the most part copying Hecataeus; there followed Democritus, Eudoxus, and others, who produced maps and geographical treatises."

With the details we are not now concerned. Viewed as a whole, however, the review is significant: every name mentioned, down to the close of the fifth century and even later, is that of an Ionian. This was natural because the Ionians at that time were far in advance of the other Greeks in their general intellectual development. Geography and the kindred subjects of history and ethnography were their special fields, and occupation with these matters became an established tradition among them. Quite naturally, therefore, the several contributors to these sciences belonged to a definite and easily discernible line of tradition. All traditions are on the whole conservative. Such

---

[270] I, 1, 11 (translation from H. L. Jones's Greek and English edition, Loeb Classical Library, London and New York, Vol. 1, 1917, p. 23).
[271] I, 1 (Carl Müller, *Geographi graeci minores,* Paris, 1882, Vol. 2, p. 471).

changes as they display in the course of their development are commonly of minor importance, the general outlines continuing to retain the essential features of the prototype. This, as we have seen, is eminently true of the cartographic tradition of the Greeks; for even in the map of Eratosthenes, which was supposed to be constructed on altogether new principles, the main features of the old Ionian maps are clearly distinguishable, though the original parallels have of course been reinterpreted in terms of degrees of latitude determined by observation of the midday height of the sun. Indeed, we may easily trace the influence of the Ionian frame in maps of a much later date; the Ionian equator is still the main axis of the charts of the fifteenth century.

In most traditions the author of the archetype is unknown. We owe it chiefly to Eratosthenes that we can name the author of the first general map of the earth. Whatever antecedent attempts in restricted areas Anaximander may have known and utilized, we may be sure that he first conceived the plan of presenting a picture of the whole earth. We may be sure also that it was he who created the frame of the map as it was to be constructed for ages. It was indeed a bold undertaking, as Eratosthenes seems to have declared. The map was of course imperfect, though capable of improvement along the lines Anaximander had sketched. And it was not to be long before the first attempt was revised and improved so that it became a thing of wonder. Anaximander is said to have died shortly after 546 B. C. Little more than a century later Hecataeus, likewise a man of Miletus, who was known for his extensive travels, made an admirable revision and extension of the map of his predecessor. Although we cannot point to many details due to his revision, we may be sure that his map was the " standard work " of the fifth and fourth centuries and therefore the one that left a permanent impress on ancient and medieval cartography.

The fact that Damastes almost slavishly copied the map is sufficient proof of this; but modern study of the geographical fragments of Hecataeus has added certainty by showing that for some regions, notably for the far west and also for the far east, he was long the sole authority. It is a far cry from Anaximander and Hecataeus in the sixth to Eratosthenes at the beginning of the second century, but despite all differences in detail, the frame of the Greek map had not changed in any of its essential features.

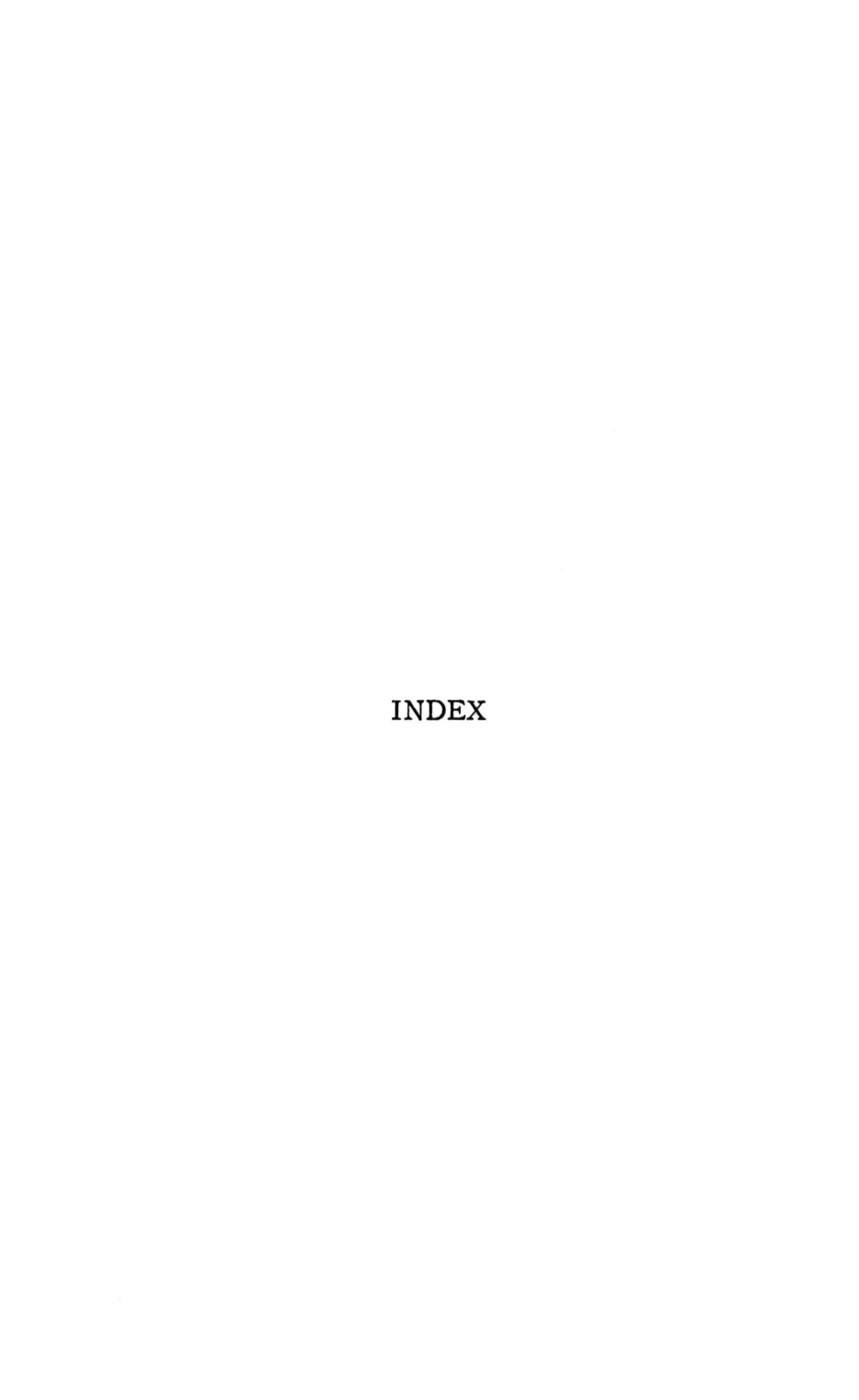

INDEX

INDEX

Achilles, shield of, 2
Adriatic Sea, 35, 36, 38
Aea, Circe's isle, 9
Aeschylus, *Prometheus Bound*, 14-16; *Prometheus Unbound*, 16-17, 37
Aëtius, epitome, 7, 11, 65, 67, 71, 72, 79, 80, 94, 100, 101
Africa, 21; circumnavigation of, 28, 57; southern tropic in, 28-30, 105-106, 124-128. *See also* Ethiopia, Nile, Syene, etc.
Agathemerus, *Outline of Geography*, 48, 99, 100, 110, 111, 132
ἀκτή, 14, 34, 112
Alexander of Aphrodisias, *In meteorologicorum libros commentaria*, 43
Alexander Polyhistor, 85
Alexander the Great, 27, 33, 49, 54, 103, 123; historians of, 13, 33, 120
Alexandria, 119, 125, 128
Amazons, 15
American Geographical Society, x
Ammon, oracle of, 23
Ammonians, 29
Anaxagoras, 68, 74, 75, 80, 81, 82
Anaximander, 57, 58, 67, 69, 72, 78, 91; map of, 11, 12, 48, 51, 57, 132-134
Anaximenes, 7, 8, 67, 69, 74, 80
Antipodes, 101
Apollonius Rhodius, *Argonautica*, 3, 35
Arabia, 13, 48
Aratus, 97-99
Archelaus, 68, 74, 77-79, 80, 82
Archimedes, 121; *Sand Reckoner*, 95, 115-117, 120
Archytas, 84, 94-95
Arctic Circle, 126, 107-108
Arganthonius, 38
Argonauts, 32, 35
Arimaspians, Scythian, 15
Aristagoras, 52
Aristarchus of Samos, 58, 120, 121
Aristophanes, *Clouds*, 70, 82, 92, 93
Aristotle, *De caelo*, 67-69, 74, 86-87; *De generatione*, 76; *Historia animalium*, 19; *Liber de inundacione Nili*, 85, 88, 101; *Metaphysics*, 68; *Meteorology*, 1, 12, 31, 39, 42-44, 78, 85, 88, 90, 96, 101; on the sphericity of the earth, 85-91, 92, 95-97, 101, 102; *Politics*, 19
Arrian, *Anabasis*, 16, 26, 35
Art of Eudoxus, 97
Asia, 11, 12. *See also* Arabia, Asia Minor, India, etc.
Asia Minor, 19, 51, 53, 54, 112
Astronomical observations, early, 8-11; geographical positions determined by, 103-109
Atarantes, 29
Atlantic coast line, 41
Atlantis, 88
Avienus, *Descriptio orbis terrae* (in Carl Müller, *Geographi graeci minores*, Paris, 1882, Vol. 2, pp. 177-189), 39, 40, 41, 113

Babylonian map, 3, 50
Bactrians, 34
Balkan peninsula, 36
Basil, St., *Hexaëmeron*, 43
Behistun, 52
Berenice, 106, 109, 119, 127
Berger, Hugo, 91, 92; *Die geographischen Fragmente des Eratosthenes*, 105, 114; *Geschichte der wissenschaftlichen Erdkunde der Griechen*, 20, 35, 37, 40, 41, 80, 95, 97, 99, 100, 106, 111, 114-117, 119, 120
Bibline Mountains, 15
Bion of Abdera, 9
Biscay, Bay of, 41
Bissing, F. W. von, *Ägyptische Weisheit und griechische Wissenschaft*, 3
Boll, Franz, *Das Lebensalter*, 70
" Brow " of Libya, 29
Bunbury, E. H., *A History of Ancient Geography*, 41, 105, 111, 120
Burnet, John, 74; *Early Greek Philosophy*, 65, 72, 77, 80, 82; edition of Plato's *Phaedo*, 83, 84; *L'expérimentation et l'observation dans la science grecque*, 94
Byzantium, 107

Callimachus, 8
Cambyses, 22
Cantor, M., *Vorlesungen über Geschichte der Mathematik*, 124

138 THE FRAME OF THE ANCIENT GREEK MAPS

Capelle, W., *Die Nilschwelle*, 102
Carthage, 128
Caspapyrus (or Caspatyrus), 18, 34
Caspian Gates, 15
Caspian Sea, 32, 51
Caucasus Mountains, 15, 31 33, 42, 43, 48, 51, 78, 113, 127
Celts, 17, 24, 35, 40, 45, 46, 47
Chalybes, 14
Choerilus, 34
Cicero, *Ad Atticum*, 110; *De natura deorum*, 67
Cilician Gates, 25, 128
Cimmerian Strait, 15
Cimmerians, 9, 47
City of the Deserters, 23, 25
Clemens Alexandrinus, *Stromata*, 3
Cleomedes, *De motu circulari corporum caelestium*, 108, 114, 117
Cleomenes, 52
Climate, 53; Asia Minor and Greece, 19, 53
Colchians, 14, 51
Colchis, 32
Cosmas, 17
Crates of Mallos, 66
Crocodiles, 27, 28
Cynetes (Cynesii, or Cynesians), 24, 37, 40, 41
Cyrene, 23
Cyrus River, 32

Damastes, 134
Darius, conquests, 13, 48, 50; tomb, 52
Deïmachus, 13, 26, 48, 49, 54, 104, 113
Democedes, 50
Democritus, 68, 69, 74, 75, 76, 100, 110, 132
Diaphragma, 43, 49, 54, 111-113, 126. See also Equator, Ionian
Dicaearchus, 13, 36, 42, 43, 49, 54, 78, 107, 110-121
Diels, Hermann, x, 85; *Doxographi graeci*, 7, 65; *Die Fragmente der Vorsokratiker*, 7, 19, 34, 70
Diodorus Siculus, *History*, 34, 101
Diogenes Laërtius, *The Lives, Doctrines, and Maxims of Famous Philosophers*, 10, 64, 67, 73, 74, 76, 79, 98
Diogenes of Apollonia, 70, 80, 83
Don River. *See* Tanaïs
Douro River, 40, 44
Doxographic tradition, 65, 79, 100

Earth, at the center of the cosmos, 86; circumference of, 88-89, 96, 97, 113-121, 123-125; dip to south, 8; disk-shaped, 7; Greek terms for, 12; hollow, 78; "outskirts," 13, 16; sphericity, 81-102, and *passim*. See also *Oikumene*
Eclipses, 80, 87, 128
Ecliptic, 71
Egypt, surveyors' or engineers' plots, 3
Egyptians, 20
Elephants, 88
Empedocles, 70
Eneti, 36
Ephorus, 16; parallelogram, 16-20, 26, 30, 34, 42, 45-47, 50
Equator, Ionian, 19, 21, 52, 53-55, 133. See also *Diaphragma*
Equinox, 59
Eratosthenes, 26, 34, 43, 48-50, 54, 66, 89, 112-119, 122-128, 131-134; calculation of the earth's circumference, 123-125; *Hermes*, 119; map, 125-128
Erythraean Sea (Indian Ocean), 27
ἐσχατιαί, 13, 16
Etearchus, 23
Ethiopia, 15, 16, 22, 23, 48
Ethiopians, 10, 17, 18, 27-29, 30, 45; Long-lived, 22, 23
"Ethiopians" of India, 47
Eudemus, 80, 103, 110, 115; *History of Astronomy*, 71
Eudoxus of Cnidus, 10, 89, 95-102, 132
Euripides, *Trojan Women*, 70
Europe, 11, 12. *See also* Celts, Ister, Scythians, etc.
Eustathius, 3
Euthymenes, 28
Euxine Sea, 31, 35

Favorinus, 73
Fobes, F. H., 44
Frame of ancient Greek maps, 3, and *passim*; eastern limit, 47-50; northern limit, 31-44; origins of, 1-25; southern limit, 26-30; western limit, 45-47
Frank, Erich, *Plato und die sogenannten Pythagoreer*, 74, 76, 78
Friedländer, P., *Die Anfänge der Erdkugelgeographie*, 101

Gades, 40
Galen, 20; *Institutio logica*, 124
Garamantes, 29

INDEX

Gaul, 39, 40
Gē, 12. See also Earth
Gisinger, Friedrich, *Die Erdbeschreibung des Eudoxos von Knidos*, 76, 100
Gnomon (sundial), 57, 58, 125
Gorgons, 15, 16
Guadalquivir River, 39

Heath, Sir Thomas, *Aristarchus of Samos*, 65, 72, 96, 98, 116
Hecataeus of Miletus, 16, 18, 19, 21, 27, 28, 33, 34, 38, 39, 40, 48, 50-52, 132-134
Heidel, W. A., *Hecataeus and the Egyptian Priests in Herodotus, Book II*, 18; *Suggestion Concerning Plato's Atlantis*, 12, 18, 88
Hemispheres, 101
Heraclitus, 23
Herakles, 16, 35, 36; shield of, 2
Herakles, Pillars of, 14, 24, 29, 49, 54, 126
Hero of Alexandria, *Dioptrics*, 123-124
Herodotus, dependence on the Ionian maps, 20-22; geographical sketch of Asia, 50-53; *History*, 2, 9-16, 18-38, 40, 42, 48, 54, 57, 59, 75, 90, 105, 112, 131; on the Ister, 24-25; on the Nile, 22-24
Herrmann, A., *Tanaïs*, 33
Hesiod, 16, 67, 73; *Scutum*, 2; *Works and Days*, 8, 59
Himalaya Mountains, 34, 47, 54, 113
Hindu Kush Mountains, 33, 42, 43, 54, 113
Hipparchus, 50, 104, 107, 115, 123, 128, 131; *In Arati et Eudoxi Phaenomena commentariorum libri tres*, 96-99, 118
Hippocrates, *De aëre*, 18-19, 22; *De flatibus*, 70; *De victu*, 59, 101
Hippocratic treatises: *On Climate, Waters, and Situations*, 19; *De septimanis*, 70
Hippolytus, *Refutatio omnium haeresium*, 7, 67, 68, 75, 77, 78
Homer, 2, 66, 67, 122; *Iliad*, 2, 8; *Odyssey*, 2, 8, 9, 27, 56, 59
Horace, *Odes*, 94
Horizon, fixed, 22, 56; designation of direction with reference to, 18, 20
Hultsch, F., *Bion aus Abdera*, 10
Hybristes River, 15
Hyperboreans, 10, 21

Iberia, 38, 46
India, geographical position, 48, 104-105; southern tropic in relation to, 26-28
Indians, 15, 17, 18, 47
Indus River, 16, 27, 43
Io, 14, 16, 27
Ionian maps, 55, 126, and *passim*; enduring frame of, 131-134
Ionians, 7, 8, 21, 83, 111; historical outlook, ix
Ister (Danube) River, 21, 22, 24-25, 31, 127; course of, 34-36; headwaters, 36-44

Jacoby, Felix, *Die Fragmente der griechischen Historiker*, 12, and *passim*; *Hekataios*, 2, 21, 33
Jaxartes River, 32
Jubilees, Book of, 33

Kiessling, A., *Hypanis*, 15

Laestrygonians, 9
Levy, Isidore, *Recherches sur les sources de la légende de Pythagore*, 73
Libya, 21
Ligurians (Ligyes), 16, 37
Longitude, 25, 127-128
Lotus Eaters, 29
Lysimachia, 114-115, 117-119

Manitius, Carl, edition of the *Phaenomena* of Hipparchus, 96-98
Marinus, 131
Martin, T. H., 59
Massalia, 37-39, 107, 126
McCrindle, J. W., translation of Cosmas, 17
Medes, 31, 51
Median axis of the Greek maps, 14. See also Equator, Ionian; Oikumene
Mediterranean basin, Greek knowledge of, 13, 37-40
Megasthenes, 104
Meissner, Bruno, *Babylonische und griechische Landkarten*, 3
Meridians, 25, 127-128
Meroë, 29, 49, 106, 125, 127
Meteors, 80
Meyer, Eduard, *Set-Typhon*, 3
Milesians, 10
Milky Way, 72
Mountains, altitudes, 111

Müllenhoff, Karl, *Deutsche Altertumskunde*, 38-41, 43, 45, 96, 97, 105, 119
Müller, Carl, *Geographi graeci minores*, 3, and *passim*
Müller, Carl and T., *Historicorum graecorum fragmenta*, 17, and *passim*
Myres, J. L., *An Attempt to Reconstruct the Maps Used by Herodotus*, 14, 18, 33, 37, 50, 53, 55
Myriandric Gulf, 51

Narbo, 46
Nasamonians, 23, 25, 29
Nearchus, 105
Necho, 28, 57
Nicagoras of Cyprus, 101
Nile River, 14, 22-24, 27, 28; first cataract, 15; floods, 21, 100-102, 111
Nilsson, Martin, *Primitive Time-Reckoning*, 59

Oceanus River, 2, 11
Oenopides of Chios, 71, 76
Oestrymnii, 40, 41
Oikumene, 17, 20, 33, 125, and *passim;* bounded by deserts, 27, 29; definition of, 12, 13, 93; dimensions, 19, 90, 100, 110; division into two parts, 22, 31, 112; median axis, 111-113, 126, 127; probable Greek concept of (map), 6. *See also* Earth
Olmstead, A. T., *History of Assyria*, 3
Onesicritus, 104
" Outskirts " of the earth (ἐσχατιαί), 13, 16. *See also* Earth

Paradise, earthly, 9
Parallels, according to Eratosthenes, 125-127; on the Ionian maps, 20
Parmenides, 11, 67, 70-76, 79, 80, 91; *Opinions of Mortals*, 72
Parnassus Mountain (Asiatic), 42
Paropanisus Mountains, 42
Persian Empire, 51
" Persian map," 13, 50, 51, 54, 112, 126
Persians, 51
Pfeiffer, Rudolfus, *Callimachi fragmenta nuper reperta*, 8
Phasis River, 14, 15, 32, 33
Philo, 105, 106, 109, 127
Philolaus, 93, 94

Phocaeans, 36, 37, 38
Phoenician chart, 33
Phoenicians, 37, 38, 41, 50, 57
Phorcides, 14
Pillars of Herakles. *See* Herakles, Pillars of
Pindar, *Isthmia*, 14, 21; *Olympica*, 21, 35
Plato, *Critias*, 19, 88; on the sphericity of the earth, 68, 81-85, 92, 94, 95, 101; *Phaedo*, 8, 68, 81, 83-85, 94; *Republic*, 8, 91; *Theaetetus*, 95; *Timaeus*, 19, 81, 84
Pliny, *Naturalis historia*, 19, 58, 105, 110, 111
Plutarch, 80
Polybius, 35, 49, 54
Posidonius of Apamea, 11, 58, 65-67, 80, 91
Posidonian areskonta, 65
Pseudo-Hippocratic treatise, *De septimanis*, 70
Pseudo-Scymnus, *Orbis descriptio* (in Carl Müller, *Geographi graeci minores*, Paris, 1882, pp. 196-237), 39, 46
Ptolemaïs Epitheras, 106
Ptolemy, Claudius, 27, 131; *Syntaxis mathematica*, 120
Ptolemy Philadelphus, 106, 119
Pygmies, 29
Pyrene, 24, 35, 39
Pyrenees Mountains, 35, 37, 39, 40, 43, 126, 127
Pythagoras, 68, 71, 72-76, 79, 84, 91
Pythagoreans, 83, 84, 92
Pytheas, 99, 106-109, 123

Red Sea, 29, 105
Rehm, A., *Antike Windrosen*, 18-19, 56
Reinhardt, Karl, *Parmenides und die Geschichte der griechischen Philosophie*, 76, 101-102
Revelation, 9
Rey, Abel, *La jeunesse de la science grecque*, 70
Rhodes, 128
Rhone River, 37
Rome, 128
Roscher, W. H., 70; *Das Alter der Weltkarte* in ' Hippokrates ' περὶ ἐβδομάδων, etc., 52
Russell, H. N., R. S. Dugan, and J. Q. Stewart, *Astronomy*, 124

INDEX

Sacae, 34
Sailing instructions, 2
Saspires, 51
Schulten, Adolf, *Tartessos,* 38
Scylax of Caryanda, voyage of, 13, 27, 34, 48, 51
Scymnus. *See* Pseudo-Scymnus
Scythia, 10, 15, 24, 31
Scythian tract, 14
Scythians, 3, 14, 16, 17, 18, 20, 33, 35, 47
Septimanis, De, 70
Sigynnae, 36
Simplicius, *Physica,* 76
Sinope, 25, 128
Skaphe, 58
Slocum, Prof. Frederick, 118
Socrates, 81-85, 92
Solstices, 57, 59
Sophocles, *Oedipus tyrannus,* 9
Spain, 16, 47
Sphericity of earth, 81-102, and *passim*
Stein, Heinrich, *Herodotos,* 23
Stephanus of Byzantium, 34
Stoics, 66, 122
Strabo, *Geography,* 1, 23, 30, 34-36, 43, 45, 46, 76, 96, 104, 105, 107, 110-112, 122; criticism of Deïmachus, 26; criticism of Eratosthenes, 66, 123; on course of Indus, 50; on early geographers, 132
Strait of Gibraltar. *See* Herakles, Pillars of
Stüve, Wilh., *Olympiodori in Aristotelis Meteora,* 42
Sundial (gnomon), 57, 58, 125
Syene, 97, 109, 114, 117-119, 124, 125, 127

Tagus River, 40, 44
Tanaïs River, 15, 32
Tannery, Paul, *Recherches sur l'histoire de l'astronomie ancienne,* 95
Tarshish, 39. *See also* Tartessus
Tartessians, 30, 38

Tartessus, 37, 39, 40, 43, 44
Tartessus River, 39
Taurus Mountains, 43, 47, 54, 112, 113, 126
Taylor, A. E., 83; *Commentary on Plato's Timaeus,* 81
Thalamas, A., *La géographie d'Ératosthène,* 119, 124, 125
Thales, 8, 67, 69, 71
Thebes, 29
Theo Smyrnaeus, *On the Mathematical Knowledge Which is Needed to Read Plato,* 71, 80, 103
Theophrastus, 73, 74, 76, 79; *Historia plantarum,* 13; *Opinions of the Physical Philosophers,* 65
Theopompus, 35
Thrace, 10, 35
Thule, 107, 126
Timosthenes, 106
Triopian headland, 51
Tropics, 103, 119, 124, and *passim;* Ionian, 21, 127; uninhabitable, 86, 90, 96. *See also* Frame of ancient Greek maps
Tyrrhenia, 38

Vetusta placita, 65
Vitruvius, *De architectura,* 58, 75
Vortex, cosmic, 7-8, 68

Webster, E. W., translation of Aristotle's *Meteorology,* 85, 96
Weissbach, F. H., *Die Keilinschriften am Grabe des Darius Hystaspis,* 52
Wellman, Max, *Eine pythagoreische Urkunde des IV. Jahrhunderts v. Chr.,* 85
Wind rose, 56
Winds, 56
Winstedt, E. O., edition of Cosmas, 17

Xenophanes, 7, 11, 91

Zeno, 66, 73, 76
Zodiac, 71
Zones, 11, 20, 53, 80, 104, 124

AMERICAN GEOGRAPHICAL SOCIETY
Research Series*

Nos. 1 and 2—*Bering's Voyages: An Account of the Efforts of the Russians to Determine the Relation of Asia and America.* By F. A. Golder. Vol. 1: *The Log Books and Official Reports of the First and Second Expeditions, 1725-1730 and 1733-1742.* With a chart of the second voyage, by Ellsworth P. Bertholf. 371 pp., 15 maps, facsimiles, etc., 1 plate, 1922. Vol. 2: *Steller's Journal of the Sea Voyage from Kamchatka to America and Return on the Second Expedition, 1741-1742.* Translated and in part annotated by Leonhard Stejneger. 291 pp., 30 maps, fascimiles, etc., 2 plates. 1925. Reprinted in 1935.

No. 3—*Battlefields of the World War, Western and Southern Fronts: A Study in Military Geography.* By Douglas Wilson Johnson. 648 pp., more than 100 photographs, 60 maps, block diagrams, and diagrams, and a separate case of plates comprising 5 maps, 3 block diagrams, and 6 panoramas. 1921.

No. 4—*The Position of Geography in British Universities.* By Sir John Scott Keltie. 33 pp. 1921.

No. 4a—*Geography in France.* By Emmanuel de Martonne. 70 pp. 1924.

No. 5—*The Agrarian Indian Communities of Highland Bolivia.* By George McCutchen McBride. 27 pp., 5 maps and photographs. 1921.

No. 6—*Recent Colonization in Chile.* By Mark Jefferson. 52 pp., 15 maps, diagrams, and photographs. 1921.

No. 7—*The Rainfall of Chile.* By Mark Jefferson. 32 pp., 10 maps and diagrams. 1922.

No. 8—*Legendary Islands of the Atlantic: A Study in Medieval Geography.* By William H. Babcock. 196 pp., 25 maps. 1922.

No. 9—*A Catalogue of Geological Maps of South America.* By Henry B. Sullivan. 191 pp., 1 map. 1922.

No. 10—*Aids to Geographical Research: Bibliographies and Periodicals.* By John Kirtland Wright. 243 pp. 1923. (Out of print.)

---

* Other series published by the Society are: *Special Publications, Library Series, Map of Hispanic America Publications,* and *Oriental Explorations and Studies.*

No. 11—*The Recession of the Last Ice Sheet in New England.* By Ernst Antevs. 120 pp., 19 maps, diagrams, and photographs, 6 plates. 1922.
No. 12—*The Land Systems of Mexico.* By George McCutchen McBride. 204 pp., 33 maps and photographs.
No. 13—*The Vegetation and Soils of Africa.* By H. L. Shantz and C. F. Marbut. 263 pp., 50 photographs, 2 maps in color in separate case. 1923. (Out of print.)
No. 14—*The Geographical Conceptions of Columbus: A Critical Consideration of Four Problems.* By George E. Nunn. 148 pp., 16 maps, 2 plates. 1924.
No. 15—*The Geographical Lore of the Time of the Crusades: A Study in the History of Medieval Science and Tradition in Western Europe.* By John Kirtland Wright. 563 pp., 12 maps and diagrams. 1925.
No. 16—*Peopling the Argentine Pampa.* By Mark Jefferson. 211 pp., 67 maps, diagrams, and photographs, 1 plate. 1926.
No. 17—*The Last Glaciation: With Special Reference to the Ice Retreat in Northeastern North America.* By Ernst Antevs. 292 pp., 30 maps and diagrams, and 9 plates. 1928.
No. 18—*The Vinland Voyages.* By Matthias Thórdarson. Translated by Thorstina Jackson Walters. 76 pp., 24 maps and photographs. 1930.
No. 19—*Chile: Land and Society.* By George McCutchen McBride. 408 pp., 58 maps, facsimiles, and photographs. 1936.
No. 20—*The Frame of the Ancient Greek Maps: With a Discussion of the Discovery of the Sphericity of the Earth.* By William Arthur Heidel. 141 pp., 1 map, 1 diagram. 1937.

HISTORY OF IDEAS
IN
ANCIENT GREECE
An Arno Press Collection

Albertelli, Pilo. **Gli Eleati:** Testimonianze E Frammenti. 1939

Allman, George Johnston. **Greek Geometry From Thales To Euclid.** 1889

Apelt, Otto. **Platonische Aufsätze.** 1912

Aristotle. **Aristotle De Anima.** With Translation, Introduction and Notes by R[obert] D[rew] Hicks. 1907

Aristotle. **Aristotle's Psychology.** With Introduction and Notes by Edwin Wallace. 1882

Aristotle. **The Politics of Aristotle.** A Revised Text With Introduction, Analysis and Commentary by Franz Susemihl and R[obert] D[rew] Hicks. 1894. Books I-V

Arnim, Hans [Friedrich August von]. **Platos Jugenddialoge Und Die Entstehungszeit Des Phaidros.** 1914

Arpe, Curt. **Das $\tau i\ \mathring{\eta}\nu\ \varepsilon \hat{\imath}\nu\alpha\iota$ Bei Aristoteles** and Hambruch, Ernst, **Logische Regeln Der Platonischen Schule In Der Aristotelischen Topik.** 1938/1904. Two vols. in one

Beauchet, Ludovic. **Histoire Du Droit Privé De La République Athénienne.** 1897. Four vols.

Boeckh, Augustus. **The Public Economy of Athens.** 1842

Daremberg, Ch[arles]. **La Médecine:** Histoire Et Doctrines. 1865

Dareste, Rodolphe [de la Chavanne]. **La Science Du Droit En Grèce:** Platon, Aristote, Théophraste. 1893

Derenne, Eudore. **Les Procès D'Impiété Intentés Aux Philosophes A Athènes Au Vme Et Au IVme Siècles Avant J. C.** 1930

Diès, A[uguste]. **Autour De Platon:** Essais De Critique Et D'Histoire. 1927

Dittmar, Heinrich. **Aischines Von Sphettos:** Studien Zur Literaturgeschichte Der Sokratiker. 1912

Dugas, L[udovic]. **L'Amitié Antique D'Après Les Moeurs Populaires Et Les Théories Des Philosophes.** 1894

Fredrich, Carl. **Hippokratische Untersuchungen.** 1899

Freeman, Kathleen. **The Work And Life Of Solon,** With A Translation Of His Poems. 1926

Frisch, Hartvig. **The Constitution Of The Athenians.** 1942

Frisch, Hartvig. **Might And Right In Antiquity.** "Dike" I: From Homer To The Persian Wars. 1949

Frutiger, Perceval. **Les Mythes De Platon:** Étude Philosophique Et Littéraire. 1930

Heidel, William Arthur. **The Frame Of The Ancient Greek Maps.** 1937

Heidel, W[illiam] A[rthur]. **Plato's Euthyphro, With Introduction and Notes and Pseudo-Platonica.** [1902]/1896. Two vols. in one

Hermann, Karl Fr[iedrich]. **Geschichte Und System Der Platonischen Philosophie.** 1839. Part One all published

Hirzel, Rudolf. **Die Person:** Begriff Und Name Derselben Im Altertum and Uxkull-Gyllenband, Woldemar Graf, **Griechische Kultur-Entstehungslehren.** 1914/1924. Two vols. in one

Kleingünther, Adolf. ΠΡΩΤΟΣ ΕΥΡΕΤΗΣ : Untersuchungen Zur Geschichte Einer Fragestellung. 1933

Krohn, A[ugust A.] **Der Platonische Staat.** 1876

Mahaffy, J. P. **Greek Life And Thought From The Age Of Alexander To The Roman Conquest.** 1887

Martin, Th[omas] Henri. **Études Sur Le Timée De Platon.** 1841. Two vols. in one

Martin, Th[omas] H[enri]. **Mémoire Sur Les Hypothèses Astronomiques.** 1879/1881. Three parts in one

Milhaud, Gaston. **Les Philosophes-Géomètres De La Grèce.** 1900

Morrow, Glenn R. **Plato's Law Of Slavery In Its Relation To Greek Law.** 1939

Plato. **The Hippias Major Attributed To Plato.** With Introductory Essay and Commentary by Dorothy Tarrant. 1928

Plato. **The Laws Of Plato.** The Text Edited With Introduction and Notes by E. B. England. 1921. Two vols.

Saunders, Trevor J. **Bibliography On Plato's Laws, 1920-1970:** With Additional Citations Through May, 1975. 1975

Plato. **The Platonic Epistles.** Translated With Introduction and Notes by J. Harward. 1932

Raeder, Hans. **Platons Philosophische Entwickelung.** 1905

Ritter, Constantin. **Neue Untersuchungen Über Platon.** 1910

Ritter, Constantin. **Platon:** Sein Leben, Seine Schriften, Seine Lehre. 1910/1923. Two vols.

Sachs, Eva. **Die Fünf Platonischen Körper.** 1917

Schwartz, Eduard. **Ethik Der Griechen.** 1951

Shute, Richard. **On The History Of The Process By Which The Aristotelian Writings Arrived At Their Present Form.** 1888

Snell, Bruno. **Die Ausdrücke Für Den Begriff Des Wissens In Der Vorplatonischen Philosophie.** 1924

Tannery, Paul. **La Géométrie Grecque.** 1887

Tannery, Paul. **Recherches Sur L'Histoire De L'Astronomie Ancienne.** 1893

Taylor, A. E. **Philosophical Studies.** 1934

Wallace, Edwin, compiler. **Outlines Of The Philosophy Of Aristotle.** 1894

Zeller, Eduard. **Platonische Studien.** 1839

Zeno And The Discovery Of Incommensurables In Greek Mathematics. 1975